# EDEXCEL
# A LEVEL MATHS
## AS LEVEL

**Authors**

Paul Hunt, Steve Cavill,
Neil Tully, Rob Wagner

# EXAM PRACTICE WORKBOOK

OXFORD
UNIVERSITY PRESS

T0202331

Great Clarendon Street, Oxford, OX2 6DP, United Kingdom

Oxford University Press is a department of the University of Oxford.

It furthers the University's objective of excellence in research, scholarship, and education by publishing worldwide. Oxford is a registered trade mark of Oxford University Press in the UK and in certain other countries

First published in 2018

British Library Cataloguing in Publication Data

Data available

978-019-841318-9

3 5 7 9 10 8 6 4

Paper used in the production of this book is a natural, recyclable product made from wood grown in sustainable forests.

The manufacturing process conforms to the environmental regulations of the country of origin.

Printed in Great Britain by Ashford Colour Press Ltd. Gosport

# Contents

## About this book

This book contains three sets of write-in, mock exam papers for the Edexcel AS Level Maths exam (7356). Full details of this exam specification can be found on the Edexcel website.

https://qualifications.pearson.com/en/qualifications/edexcel-a-levels/mathematics-2017.html

There are two papers in each exam set: Paper 1 covers Pure Mathematics and Paper 2 covers Statistics and Mechanics. Paper 1 is 120 minutes long and is worth 100 marks. Paper 2 is 75 minutes long and is worth 60 marks.

### The Large data set

The AS Level examination will assume that you are familiar with a Large data set (LDS). In the exam, some questions will be based on the LDS and may include extracts from it. It is Edexcel's intention that you should be taught using the LDS as this will give you a material advantage in the exam.

The data set consists of weather data samples provided by the Met Office for five UK weather stations and three overseas weather stations in the time periods May to October 1987 and May to October 2015. An Excel spreadsheet containing the LDS is available from the Edexcel website at the address given above.

### Answers

The back of this book contains short answers to all the questions.
Full mark schemes for each mock paper can be found online.

https://global.oup.com/education/content/secondary/series/edexcelalevelmaths-answers

### Formulae

In the exam, you will be provided with a 'Mathematical formulae and statistics tables' booklet that is for use in AS Level and A Level Maths qualifications. The relevant AS Level Maths formulae and statistical tables are provided at the end of this book.

### Calculators

All papers are calculator papers. You must make sure that you have a calculator and that you know how to use it. The rules on which calculators are allowed can be found in the Joint Council for General Qualifications document 'Instructions for conducting examinations' (ICE).

# AS Level Mathematics

## *Paper 1 (Set A)*

# Edexcel

| Name | | Class | |
| Signature | | Date | |

## Materials

Any calculator permitted by ICE regulations may be used. Calculators must not have retrievable mathematical formulae stored in them and they must not be able to differentiate, integrate or manipulate algebra.

## Instructions

- Use a black or blue pen.
- Use a dark pencil if not using a pen, for diagrams/graphs.
- Write your name and class at the top of this page in the box provided.
- Answer all the questions in this paper.
- Answer the questions in the spaces given.
- You may not need to use all the space given.
- You should make your methods clear by showing sufficient working or you may not gain full credit for your answers.
- Give inexact answers to three significant figures, unless otherwise stated.

| Question | Mark |
|----------|------|
| 1 | |
| 2 | |
| 3 | |
| 4 | |
| 5 | |
| 6 | |
| 7 | |
| 8 | |
| 9 | |
| 10 | |
| 11 | |
| 12 | |
| 13 | |
| Total | |

## Information

- Extracts from the booklet 'Mathematical Formulae and Statistical Tables' have been provided.
- This question paper has 13 questions.
- There are 100 marks in total for this paper.
- The marks available for each question are in brackets.
- You should aim to spend approximately one minute per mark on each question.

## Advice

- Read each question carefully before you begin your answer.
- Attempt to answer all the questions.
- If you have time at the end, check your answers.
- If you start an answer again, cross out the old workings.

Answer **all** questions in the spaces provided.

**1** **a** Integrate these expressions with respect to $x$

    **i**   $x^3$                                                                   **[1 mark]**

    **ii**   $x^{-2}$                                                                  **[1 mark]**

    **iii**   $\dfrac{1}{x^5}$                                                               **[1 mark]**

  **b**   Hence, or otherwise, find $\displaystyle\int \left( x^2 + \frac{1}{x} \right)\left( \frac{3}{x^4} - 2x \right)\mathrm{d}x$            **[4 marks]**

**2  a**    Use the factor theorem to show that $(x-6)$ and $(x+1)$ are both factors of the polynomial $f(x)$

$$f(x) = x^4 - x^3 - 22x^2 - 44x - 24$$    **[3 marks]**

**b**    Hence factorise $f(x)$ completely.    **[4 marks]**

**3**     In the diagram shown below, AB = AC = 10 cm, CE = 12 cm and DE = 5 cm.

Angle ABC = 75°

Find the ratio of the area of triangle ABC to the area of triangle ADE.

Express your answer in the form 1 : $n$                                   [5 marks]

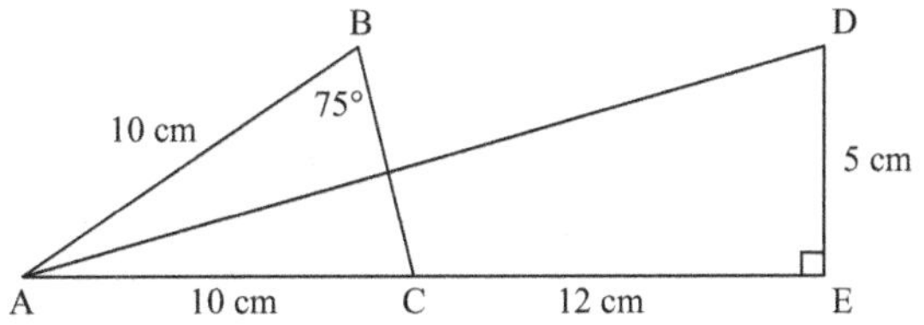

**4**     Find the exact solution(s) to this equation.

$$\sqrt{2x}\,\sqrt{x+2} = 3$$

[4 marks]

**5** Write this expression $\frac{1}{2}\log_{10} b - \log_{10} 1 - \frac{1}{4}\log_{10} c^2$ as a single logarithm with a coefficient of 1 **[4 marks]**

**6** What is the coefficient of $x^{10}$ in the binomial expansion of $\left( x^2 - \frac{1}{4} \right)^8$ ? **[2 marks]**

**7** Use differentiation from first principles to find the derivative of this function
$q(x) = 4x^2 - 2x$ with respect to $x$ **[5 marks]**

AS Level Mathematics Paper 1 (Set A)

**8**  A circle C has equation $x^2 + y^2 - 6x - 12y + 41 = 0$

**a**  By completing the square, find the radius and coordinates of the centre of circle C.    **[4 marks]**

**b**  The circle C is a translation of the circle with equation $x^2 + y^2 = 4$

Describe this transformation geometrically.    **[2 marks]**

**8  c  i**    The points A (1, 6) and B (3, 8) lie on circle C.

Find the equation of the perpendicular bisector of the line AB in the form $y = px + q$     **[6 marks]**

**ii**    Find the exact coordinates of both points where the perpendicular bisector
found in part **c i** intersects circle C.                                            **[6 marks]**

**9** Solve these equations for all values of $\theta$ in the interval $0° \leq \theta \leq 90°$

**a** $\cos^2 \theta + \sin \theta = 1$ [4 marks]

**b** $\tan^2 5\theta = 1$ [3 marks]

**10** A curve has equation $y = 3x^2 + \dfrac{3}{x^2}$

Find the area between the curve, the $x$-axis and the lines $y = a$ and $y = 3a$ where $a$ is a positive constant. Give your answer in terms of $a$ [4 marks]

**11**    It is claimed that the data in the table is consistent with a relationship of the form $y = ab^x$

| $x$ | 1 | 5 | 7 | 8 | 12 |
|---|---|---|---|---|---|
| $y$ | 1.67 | 2.19 | 2.51 | 2.68 | 3.51 |
| $\log_{10} y$ | | | | | |

**a  i**   Show that this relationship can also be written in the linear form

$\log_{10} y = x \log_{10} b + \log_{10} a$                                                                **[3 marks]**

_____

_____

_____

_____

_____

_____

_____

_____

**ii**   By completing the table above and then drawing a suitable graph, explain whether or not this
claim is true.                                                                                    **[3 marks]**

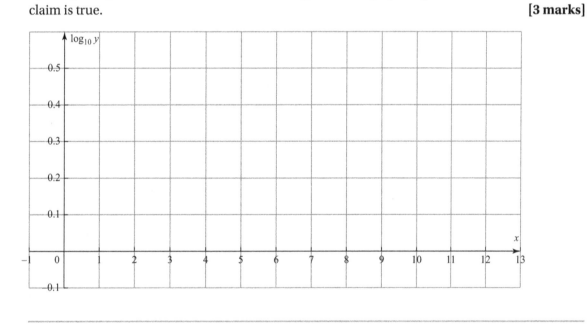

_____

_____

_____

_____

_____

_____

_____

**11 b** Estimate the values of the constants $a$ and $b$

Give your answers to 2 significant figures.

[5 marks]

**12**    The circles $C_1$ and $C_2$ share a common chord AB.

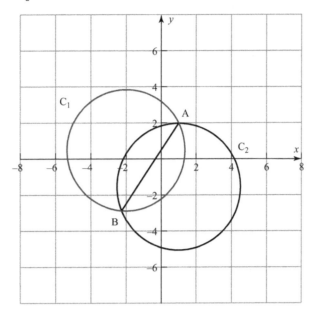

The equations of the circles are

$$C_1:\ 4(x+2)^2+4\left(y-\frac{1}{2}\right)^2=45$$

$$C_2:\ 4(x-1)^2+4\left(y+\frac{3}{2}\right)^2=49$$

**a    i**    Find the equation of the chord AB in the form $y = ax + b$                    **[4 marks]**

**12 a ii** Hence prove that the coordinates of point A are (1, 2) and point B are $\left(-\dfrac{29}{13}, -\dfrac{37}{13}\right)$     **[5 marks]**

**b** Find the distance between the points A and B.
Give your answer to 3 decimal places.     **[2 marks]**

**13** A cuboid has a base of length $4x$ cm and width $3x$ cm.

The combined total of all edges of the cuboid is 16 cm.

**a** Find an expression for the height of the cuboid in terms of $x$ [2 marks]

**b** Show that the surface area of the cuboid is $56x - 74x^2$ [3 marks]

**13 c i** Find the exact value of the maximum possible surface area of the cuboid. [7 marks]

**ii** State the exact dimensions of the cuboid when the surface area is at its maximum. [3 marks]

**End of questions**

# AS Level Mathematics    Edexcel
## Paper 2 (Set A)

| Name | | Class | |
|---|---|---|---|
| Signature | | Date | |

## Materials

Any calculator permitted by ICE regulations may be used. Calculators must not have retrievable mathematical formulae stored in them and they must not be able to differentiate, integrate or manipulate algebra.

## Instructions

- Use a black or blue pen.
- Use a dark pencil if not using a pen, for diagrams/graphs.
- Write your name and class at the top of this page in the box provided.
- Answer all the questions in this paper.
- Answer the questions in the spaces given.
- You may not need to use all the space given.
- You should make your methods clear by showing sufficient working or you may not gain full credit for your answers.
- Give inexact answers to three significant figures, unless otherwise stated.

## Information

- Extracts from the booklet 'Mathematical Formulae and Statistical Tables' have been provided.
- This question paper has 10 questions.
- There are 60 marks in total for this paper.
- The marks available for each question are in brackets.
- You should aim to spend approximately one minute per mark on each question.

## Advice

- Read each question carefully before you begin your answer.
- Attempt to answer all the questions.
- If you have time at the end, check your answers.
- If you start an answer again then cross out the old workings.

| Question | Mark |
|---|---|
| 1 | |
| 2 | |
| 3 | |
| 4 | |
| 5 | |
| 6 | |
| 7 | |
| 8 | |
| 9 | |
| 10 | |
| Total | |

# Section A

Answer **all** questions in the spaces provided.

**1**    The daily mean air temperatures, $T\,^\circ C$, over a two-week period at the start of June are measured for a city.

27.3    26.2    26.6    19.2    25.4    22.4    21.1    24.6    24.2    20.8    23.7    25.7    23.1    26.6

**a**    For this data calculate

   **i**    The mean,                                                                                            [1 mark]

   **ii**    The standard deviation.                                                                    [1 mark]

**b**    Due to a labelling error it isn't known whether the data is for Perth or Beijing.
State which city the data is for, explaining your reasoning.                    [1 mark]

**c**    Temperatures in Fahrenheit are given by $T\,(^\circ F) = 1.8\,T\,(^\circ C) + 32$
Using Fahrenheit calculate

   **i**    The mean, $\mu$,                                                                                    [1 mark]

**1  c  ii**  The standard deviation, σ                                                    **[1 mark]**

_____

_____

_____

_____

**2**  As part of a large weather conditions survey, data are collected on the average amount of rainfall measured per week by weather stations in the UK. In 1987 a weather station measured an average of 11.1 mm of rainfall per week. A newspaper takes a sample from a local weather station in 2015 to try to identify any changes in the amount of rainfall per week.

**a**  The newspaper only takes a sample from its own town, claiming that it doesn't expect weather in that town to differ from those of the whole UK and so it doesn't need to take a sample from other locations.

Give the name of this sampling method.                                          **[1 mark]**

_____

_____

_____

_____

_____

**b**  The newspaper's sample data are grouped in this table showing the amount of rainfall per week to the nearest 1 mm and the number of weeks that rainfall was at that level. There are 44 weeks in the sample.

| Weekly rainfall (mm) | ≤ 6 | 7 | 8 | 9 | 10 | 11 | 12 | 13 | 14 | 15 | ≥ 16 |
|---|---|---|---|---|---|---|---|---|---|---|---|
| Number of weeks | 5 | 3 | 4 | 10 | 7 | 3 | 5 | 4 | 1 | 2 | 0 |

Estimate the probability that a randomly-chosen week in this sample experienced at least an average of 11.5 mm of rainfall.                                          **[2 marks]**

_____

_____

_____

_____

_____

_____

_____

**2  c**    Data is also collected on the total number of hours of sunshine on average per week.
It is found that the number of hours of sunshine is independent of the amount of rainfall.

Explain what it means for these quantities to be independent.    **[1 mark]**

_____

_____

_____

_____

_____

_____

_____

_____

**3**    The percentage score that each student in a class achieves on their Maths and English exams are shown in the scatter diagram.

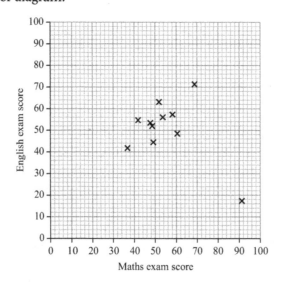

**a**    Circle any outlier(s) in the data, and explain why you believe they are outliers.    **[2 marks]**

_____

_____

_____

_____

_____

_____

_____

_____

**3 b** Ignoring the outliers you identified in part **a**, add a line of best fit to the
scatter graph. **[1 mark]**

**c** One student scored 65% on the Maths exam but was ill for the English exam, so their
data has been omitted from the scatter diagram.

Using your line of best fit, estimate the score they might have achieved on
the English exam. Show how you determined your estimate. **[2 marks]**

_____

_____

_____

_____

_____

_____

_____

_____

_____

_____

_____

_____

_____

**d** Using the data from the scatter diagram, fill in this grouped frequency table. **[3 marks]**

| Maths exam score | $0 \leq x < 40$ | $40 \leq x < 45$ | $45 \leq x < 50$ | $50 \leq x < 55$ | $55 \leq x < 60$ | $60 \leq x < 70$ | $70 \leq x < 100$ |
|---|---|---|---|---|---|---|---|
| Frequency | | | | | | | |

**e** A student is randomly chosen from the class.

**i** Calculate the probability that they scored at least 45% but less than 60%
on the Maths exam. **[2 marks]**

_____

_____

_____

_____

_____

_____

_____

_____

_____

**3 e ii** Given they scored at least 45% but less than 60% on the Maths exam, calculate the probability that they scored at least 50%. **[3 marks]**

**4** In a weather station factory there is a fixed probability of 6% that any one station isn't fit for sale, independent of any other station. The factory manager believes that the fixed probability of a station not being fit for sale has increased. To test this hypothesis the manager wants to take a sample of stations produced in the factory during one week.

**a i** The factory manager wants to use simple random sampling to take a sample of size 18 from the 3000 stations produced that week.

Describe how the manager can take a simple random sample. **[3 marks]**

**ii** Explain why this sample is not expected to be biased. **[1 mark]**

**4    b**    In the sample of 18 stations, three of them are not fit for sale.

Carry out a hypothesis test at the 5% significance level to investigate if the manager's belief is supported by the evidence.                                                    **[4 marks]**

**End of section A**

# Section B

Answer **all** questions in the spaces provided.

**5**  Usain Bolt ran his fastest 100 m race in 9.58 s, giving an average speed of 10.44 m s$^{-1}$

What was his average speed in kilometres per hour?  **[1 mark]**

**6**  The velocity-time graph represents the motion of a car.

Calculate the distance travelled by the car during these 17 seconds.  **[2 marks]**

**7** The velocity, $v\,\mathrm{ms}^{-1}$, of a body at time $t$ seconds is given by the formula

$$v = 4t^2 - 6t + 3$$

Find a formula for the displacement of the body at time $t$ given that the initial displacement is 5 m.  **[3 marks]**

**8** A force of magnitude 37 N has a component of 35 N in the $x$-direction.

**a** Calculate the component of the force in the $y$-direction.  **[3 marks]**

**8  b**    Find the angle between the resultant force and the *x*-direction.                          **[3 marks]**

**9**       In this question take *g* as $10 \text{ m s}^{-2}$

A particle is projected vertically downwards from point P with velocity $6 \text{ m s}^{-1}$

It reaches the ground after 3 s.

**a**    Calculate the height of point P above the ground.                          **[2 marks]**

**b**    Calculate the speed of the particle at the instant it hits the ground.                          **[2 marks]**

**9** **c** Calculate the time taken from the point of projection until the particle is 10 m above the ground. **[5 marks]**

**d** Explain how your answer to part **c** would change if you included the effect of air resistance on the particle. **[2 marks]**

**10** In this question take $g$ as $10 \text{ m s}^{-2}$

Two masses of 5 kg and 8 kg are connected by a light, inextensible string that passes over a smooth, frictionless pulley. The masses are held stationary.

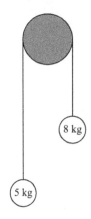

**a** The masses are released from rest and accelerate at $a \text{ m s}^{-2}$
The tension in the string is $T$ N.

By considering the forces acting on the 8 kg mass, show that $80 - T = 8a$ **[2 marks]**

**b** By first finding another equation, calculate both $T$ and $a$ **[5 marks]**

**End of questions**

# AS Level Mathematics
## Paper 1 (Set B)

# Edexcel

| Name | | Class | |
|------|--|-------|--|
| Signature | | Date | |

## Materials

Any calculator permitted by ICE regulations may be used. Calculators must not have retrievable mathematical formulae stored in them and they must not be able to differentiate, integrate or manipulate algebra.

## Instructions

- Use a black or blue pen.
- Use a dark pencil if not using a pen, for diagrams/graphs.
- Write your name and class at the top of this page in the box provided.
- Answer all the questions in this paper.
- Answer the questions in the spaces given.
- You may not need to use all the space given.
- You should make your methods clear by showing sufficient working or you may not gain full credit for your answers.
- Give inexact answers to three significant figures, unless otherwise stated.

| Question | Mark |
|----------|------|
| 1 | |
| 2 | |
| 3 | |
| 4 | |
| 5 | |
| 6 | |
| 7 | |
| 8 | |
| 9 | |
| 10 | |
| 11 | |
| 12 | |
| 13 | |
| **Total** | |

## Information

- Extracts from the booklet 'Mathematical Formulae and Statistical Tables' have been provided.
- This question paper has 13 questions.
- There are 100 marks in total for this paper.
- The marks available for each question are in brackets.
- You should aim to spend approximately one minute per mark on each question.

## Advice

- Read each question carefully before you begin your answer.
- Attempt to answer all the questions.
- If you have time at the end, check your answers.
- If you start an answer again then cross out the old workings.

Answer **all** questions in the spaces provided.

1      Use the fact that $\log_{10} 3 = 0.47712$ to 5 decimal places to calculate the following to 4 decimal places. You must show your working.

**a**      $\log_{10} 27$            **[3 marks]**

**b**      $\log_{10} 300$            **[3 marks]**

**c**      $\log_{10}\left(\dfrac{300}{27}\right)$            **[2 marks]**

**2**   Sketch each of the following graphs on the separate sets of axes provided.

Make sure that you clearly state the equations of any asymptotes, and
the coordinates of all points where the graphs intersect the coordinate axes.

**a**   $y - 2 = (x + 3)^2$                                                    [3 marks]

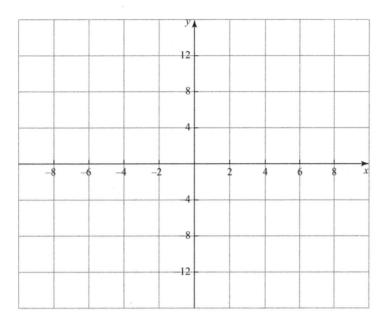

**2    b**    $y = (4-x)(2x-3)^2$                                                                 **[3 marks]**

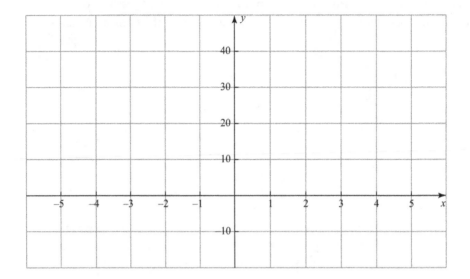

**c**    $y = \dfrac{1}{2(x+5)^2}$                                                                **[3 marks]**

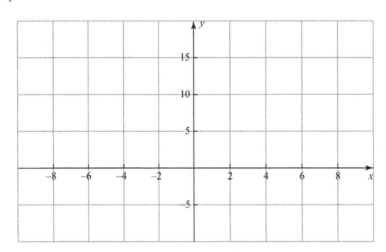

**2    d**    $y = 2 \sin 2x$ [3 marks]

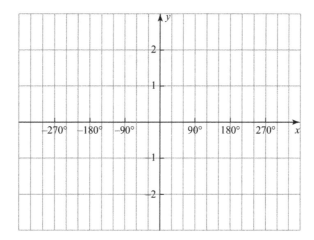

_____

_____

_____

_____

_____

_____

_____

_____

**3**    Prove that $t^2 - t$ is even for all positive integers, $t$ [4 marks]

_____

_____

_____

_____

_____

_____

_____

_____

_____

_____

_____

_____

**4** James says he has solved the equation below but Linda claims that his solution is wrong.

Who is correct? Explain your answer.

$$x^2 - 6x + 8 = 3$$
$$(x-2)(x-4) = 3$$
$$x-2 = 3 \text{ or } x-4 = 3$$
$$x = 5 \text{ or } x = 7 \qquad \text{[1 mark]}$$

**5**    Solve this pair of simultaneous equations.

$$2y^2 - x^2 + xy = 14$$

$$5x + 4y = 7$$    **[6 marks]**

**6** Solve this equation, to 1 decimal place, for values of $\theta$ in the interval $0 < \theta < 360°$

$$4\cos^2\theta + \sin\theta = 2$$

[5 marks]

**7**   A curve C and a line L have equations $y = 6x^2 + 8x - 4$ and $y = p(2x-1)$

**a**   Show that the $x$-coordinate of any point of intersection of C and L satisfies the equation

$$6x^2 + 2(4-p)x + p - 4 = 0$$

**[1 mark]**

**b**   C and L intersect at two distinct points.

**i**   Show that $p^2 - 14p + 40 > 0$

**[3 marks]**

**ii**   Find the possible values of $p$

**[4 marks]**

**8**  Two sides of a parallelogram are 6.6 cm and 8.9 cm long.

One of the diagonals of the parallelogram is 5.7 cm long.

What is the area of the parallelogram?

Give your answer to one decimal place.                    **[4 marks]**

**9** The three points A, B and C have positions vectors

$$\overrightarrow{OA} = \mathbf{i} + \mathbf{j}, \quad \overrightarrow{OB} = 3\mathbf{i} - 2\mathbf{j}, \quad \overrightarrow{OC} = 7\mathbf{i} - 8\mathbf{j}$$

**a** Find a unit vector in the direction of $\overrightarrow{AC}$                                            **[3 marks]**

**b** Find the resultant vector, $\overrightarrow{OD}$, of $\overrightarrow{OA}$ and $\overrightarrow{OB}$, and calculate the exact magnitude and bearing of $\overrightarrow{OD}$                                            **[6 marks]**

**9 c i** By finding vectors $\overrightarrow{AB}$ and $\overrightarrow{BC}$ prove that A, B and C are collinear. [5 marks]

**ii** What multiple of $\overrightarrow{AC}$ is $\overrightarrow{BC}$? [1 mark]

**10** The value, $V$, of a particular mobile phone depends on $t$, the age of the phone in years from the date of purchase, and is given by the formula below.

$$V = 150\left(1 + 4e^{-\frac{t}{3}}\right)$$

**a** What was the purchase price of the phone? [1 mark]

**b** How much will the phone be worth two years after it was purchased? [1 mark]

**c** How long after it was first purchased will the phone have lost half of its initial value? [4 marks]

**10 d** Sketch a graph of $V$ against $t$, showing how the value of the phone varies over time. **[4 marks]**

**e** Do you think the model is realistic? Explain your answer. **[1 mark]**

**11**  The diagram shows a kite, whose diagonals are of length $a$ cm and $(4+\sqrt{3})$ cm.

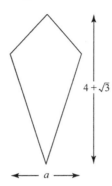

$4+\sqrt{3}$

$a$

The area of the kite is $\left(\dfrac{5}{2}-\sqrt{3}\right)$ cm$^2$.

Calculate the *exact* length $a$. You must show your workings.  **[5 marks]**

**12** The graph below shows a quadratic curve intersecting the $x$-axis at the points $(a, 0)$ and $(1, 0)$ and the $y$-axis at the point $(0, 1)$

The graph has its vertex at the point $\left(\dfrac{1}{4}, \dfrac{9}{8}\right)$

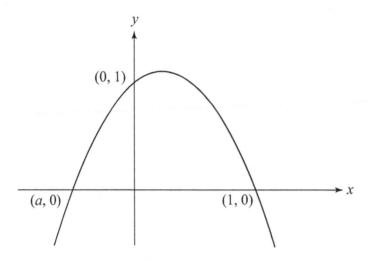

**a** Find the equation of the quadratic curve. **[7 marks]**

**12 b** By factorising the equation found in part **a**, find the value of $a$ [2 marks]

**c** Find the area bounded by the curve and the $x$-axis.

You must show your workings. [5 marks]

**13** The second derivative of a function, $u(x)$, is given by $\dfrac{d^2u}{dx^2} = 30x + 8$

When $x = -1$, $\dfrac{du}{dx} = 7$

**a** Find an expression for $\dfrac{du}{dx}$ in terms of $x$ [4 marks]

**b** When $x = -1$, $u = -2$

Find an expression for $u$ in terms of $x$ [3 marks]

**End of questions**

# AS Level Mathematics

**Edexcel**

*Paper 2 (Set B)*

| Name | | Class | |
| --- | --- | --- | --- |
| Signature | | Date | |

## Materials

Any calculator permitted by ICE regulations may be used. Calculators must not have retrievable mathematical formulae stored in them and they must not be able to differentiate, integrate or manipulate algebra.

| Question | Mark |
| --- | --- |
| 1 | |
| 2 | |
| 3 | |
| 4 | |
| 5 | |
| 6 | |
| 7 | |
| 8 | |
| Total | |

## Instructions

- Use a black or blue pen.
- Use a dark pencil if not using a pen, for diagrams/graphs.
- Write your name and class at the top of this page in the box provided.
- Answer all the questions in this paper.
- Answer the questions in the spaces given.
- You may not need to use all the space given.
- You should make your methods clear by showing sufficient working or you may not gain full credit for your answers.
- Give inexact answers to three significant figures, unless otherwise stated.

## Information

- Extracts from the booklet 'Mathematical Formulae and Statistical Tables' have been provided.
- This question paper has 8 questions.
- There are 60 marks in total for this paper.
- The marks available for each question are in brackets.
- You should aim to spend approximately one minute per mark on each question.

## Advice

- Read each question carefully before you begin your answer.
- Attempt to answer all the questions.
- If you have time at the end, check your answers.
- If you start an answer again then cross out the old workings.

# Section A

Answer **all** questions in the spaces provided.

**1**     The probability that an umbrella sold at a second-hand shop has a hole in it is 0.15

The probability that one of the umbrellas has a broken handle is 0.05

The probability of having a hole and the probability of having a broken handle are independent.

**a**     Find the probability that a randomly chosen umbrella has a hole in it and has a
broken handle.                                                                                          **[1 mark]**

**b**     Two umbrellas are chosen at random from different second-hand shops.

**i**     Find the probability that exactly one umbrella has a hole in it.          **[2 marks]**

**ii**    Find the probability that at least one umbrella has a hole in it and at least one
umbrella has a broken handle.                                                              **[2 marks]**

**2**   Data is collected, using opportunity sampling, and the following results are obtained.

   2.3   2.6   2.7   3.5   3.8   4.1   4.2   4.5   5.5   6.8

**a** **i** Explain what is meant by opportunity sampling. **[1 mark]**

_____

_____

_____

**ii** Calculate the mean and standard deviation of this sample. **[2 marks]**

_____

_____

_____

_____

_____

_____

**b** Another sample is taken using systematic sampling.
That sample has $N = 80$, $\Sigma x = 256$ and $\Sigma x^2 = 859$

**i** Give one reason why opportunity sampling is more likely than systematic
sampling to provide a biased sample. **[1 mark]**

_____

_____

_____

**ii** Calculate the mean and standard deviation of the second sample. **[3 marks]**

_____

_____

_____

_____

_____

**iii** Compare and contrast the two samples using the information given. **[2 marks]**

_____

_____

_____

_____

_____

3  The amount of rainfall per day in the countryside is monitored over time and presented in this histogram.

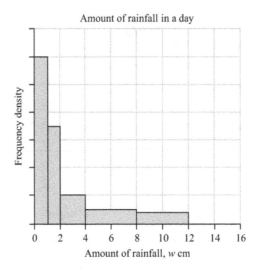

Amount of rainfall in a day

Frequency density

Amount of rainfall, $w$ cm

a  Use the histogram to complete this probability distribution table.  **[4 marks]**

| Amount of rainfall | $0 \le w < 1$ | $1 \le w < 2$ | $2 \le w < 4$ | $4 \le w < 8$ | $8 \le w < 12$ | $12 \le w$ |
|---|---|---|---|---|---|---|
| Frequency | 12 | | | | | |

b  Use your table in part **a** to estimate the probability that at least 6 cm of rain falls on a random day.  **[2 marks]**

**3 c** A weather scientist thinks that the sunny weather forecast over the next two weeks is likely to reduce the amount of rainfall. They look to see if there are significantly fewer days with six or more cm of rainfall.

**i** State two conditions required for a binomial distribution to be an appropriate model and explain why they apply to this situation. **[4 marks]**

**ii** Over the two sunny weeks there were no days on which 6 cm or more of rain fell.

By calculating the p-value of the result, perform a hypothesis test at the 10% level to see if there is less rainfall during the sunny weeks.
State your hypotheses clearly. **[6 marks]**

**End of section A**

# Section B

Answer **all** questions in the spaces provided.

**4**  A particle moves with constant acceleration of 4 m s⁻²

It accelerates from 2 m s⁻¹ to 10 m s⁻¹

Calculate the distance it travels while accelerating.                    **[2 marks]**

**5**  This displacement-time graph models the motion of a bicycle.

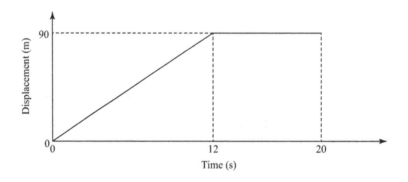

**a**  Describe what the bicycle is doing from $t = 12$ s to $t = 20$ s          **[1 mark]**

**5 b** Calculate the velocity of the bicycle during the first 12 seconds. **[2 marks]**

**c** Ian decides to improve the model so that the bicycle is *accelerating* for the first 6 seconds then *decelerating* for the next 6 seconds.

Sketch the displacement-time graph for the first 12 seconds of Ian's model. **[2 marks]**

**6** A particle moves such that its velocity, $v$ m s$^{-1}$, after time, $t$ s, is given by
$v = 12t - 3t^2$

**a** Find an expression for the acceleration of the particle after $t$ seconds. **[2 marks]**

**6 b** The initial displacement of the particle is 5 m.

Show that the displacement when $t = 4$ s is 37 m. **[5 marks]**

**c** Find the times when the particle is at rest. **[3 marks]**

**7** A car of mass 800 kg is pulling a trailer of mass 200 kg.

The car and trailer experience resistances of 100 N and 250 N respectively.
The car exerts a driving force of 500 N.

**a** Find the acceleration of the car and trailer. **[3 marks]**

**7    b**    Find the tension in the tow bar.                                                    **[3 marks]**

**8**    A model rocket, with mass 10 kg and starting from rest, is fired from the top of a building by a giant slingshot.

The slingshot accelerates the rocket upwards to a speed of 35 m s$^{-1}$, over a time period of 1 s, before releasing it.

Assuming no air resistance, calculate how high the rocket reaches above its initial starting position (at rest), before falling back towards the ground.                **[7 marks]**

**End of questions**

# AS Level Mathematics
## *Paper 1 (Set C)*

# Edexcel

| Name | | Class | |
|---|---|---|---|
| Signature | | Date | |

## Materials

Any calculator permitted by ICE regulations may be used. Calculators must not have retrievable mathematical formulae stored in them and they must not be able to differentiate, integrate or manipulate algebra.

## Instructions

- Use a black or blue pen.
- Use a dark pencil if not using a pen, for diagrams/graphs.
- Write your name and class at the top of this page in the box provided.
- Answer all the questions in this paper.
- Answer the questions in the spaces given.
- You may not need to use all the space given.
- You should make your methods clear by showing sufficient working or you may not gain full credit for your answers.
- Give inexact answers to three significant figures, unless otherwise stated.

| Question | Mark |
|---|---|
| 1 | |
| 2 | |
| 3 | |
| 4 | |
| 5 | |
| 6 | |
| 7 | |
| 8 | |
| 9 | |
| 10 | |
| 11 | |
| 12 | |
| Total | |

## Information

- Extracts from the booklet 'Mathematical Formulae and Statistical Tables' have been provided.
- This question paper has 12 questions.
- There are 100 marks in total for this paper.
- The marks available for each question are in brackets.
- You should aim to spend approximately one minute per mark on each question.

## Advice

- Read each question carefully before you begin your answer.
- Attempt to answer all the questions.
- If you have time at the end, check your answers.
- If you start an answer again then cross out the old workings.

Answer **all** questions in the spaces provided.

**1**  **a**  There are no prime numbers between 115 and 125 inclusive.

Is this statement correct?

Prove or disprove your answer.  **[3 marks]**

_____

_____

_____

_____

_____

_____

_____

_____

_____

_____

_____

_____

_____

_____

_____

_____

**b**  If the product $ab$ is a rational number, then the numbers $a$ and $b$ are also rational numbers.

Disprove this statement.  **[1 mark]**

_____

_____

_____

_____

_____

_____

_____

_____

_____

_____

_____

**2**  **a**  A point on the unit circle has coordinates $\left(-\dfrac{\sqrt{2}}{2}, \dfrac{\sqrt{2}}{2}\right)$

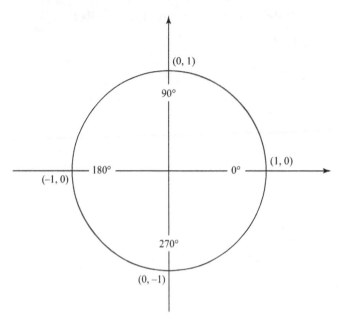

What obtuse angle does the point represent?                                      **[1 mark]**

**b**  $\cos\alpha = \dfrac{15}{17}$, where $\alpha$ is a reflex angle.

**i**  Derive an exact value for $\sin\alpha$                                      **[4 marks]**

**2 b ii** Derive an exact value for tan $\alpha$ [2 marks]

**c** Prove this trigonometric identity.

$$\frac{\cos x}{1-\sin x} \equiv \frac{1+\sin x}{\cos x}$$

[4 marks]

**3**    Solve these inequalities exactly.

**a**
$$-2 \leq \frac{6-3x}{5} < 6$$
[3 marks]

**b**
$$4(2-x) - 3(8-3x) < 4 - 3(2x+1) + x$$
[3 marks]

**3** **c** Habib's teacher has challenged him to solve the inequality $\frac{x-6}{x+4} < 6$

Habib's attempt is shown below.

$$\frac{x-6}{x+4} < 6$$

$$x - 6 < 6(x + 4)$$

$$x - 6 < 6x + 24$$

$$5x > -30$$

$$x > -6$$

Is Habib's solution correct?

Justify your answer. **[2 marks]**

_____

_____

_____

_____

_____

_____

_____

_____

**4** Daryl has solved the equation $3 \sin^2 x = \sin x$, where $0° \leq x \leq 180°$

His solution is shown below.

$$3 \sin^2 x = \sin x$$

$$3 \sin x = 1$$

$$\sin x = \frac{1}{3}$$

$$x = 19.5°, \ x = 160.5° \ (1 \text{ dp})$$

**a** Why is Daryl's solution incorrect? **[1 mark]**

_____

_____

_____

_____

_____

_____

_____

_____

**4  b**    Solve the equation correctly.                                    **[3 marks]**

**5**    Two consecutive, positive, even numbers are chosen.

The square of both numbers is taken and then the smaller answer is multiplied by the reciprocal of the larger answer.

This gives a final value of $\dfrac{81}{100}$

What are the two consecutive even numbers?                                    **[6 marks]**

**6** A curve has equation $y = 3x^4 + 18x^3 - 15x^2 + 6$

**a  i**  Find $\dfrac{dy}{dx}$                                                  [2 marks]

**ii**  Find $\dfrac{d^2y}{dx^2}$                                                [1 mark]

**b**  The point T, with coordinates (1, 12), lies on the curve.

Determine whether $y$ is increasing or decreasing at point T.

Give a reason for your answer.                                  [2 marks]

**6    c**    Find the coordinates of the three stationary points of $y$.

In each case, determine the nature of the stationary point.                    **[8 marks]**

**7  a**  It is claimed that the two variables, $x$ and $y$, are connected by a relationship of the form $y = kx^n$ where $k$ and $n$ are constants.

| $x$ | 3 | 4 | 6 | 9 | 10 |
|---|---|---|---|---|---|
| $y$ | 14.50 | 22.97 | 43.95 | 84.09 | 99.53 |
| $\log_{10}x$ | | | | | |
| $\log_{10}y$ | | | | | |

**i**  Show that this relationship can also be written in the linear form

$$\log_{10} y = \log_{10} k + n\log_{10} x$$

[3 marks]

_____

_____

_____

_____

_____

_____

_____

**ii**  By completing the table of $x$ and $y$ values and then drawing a suitable straight-line graph, test whether or not this claim is true.

[4 marks]

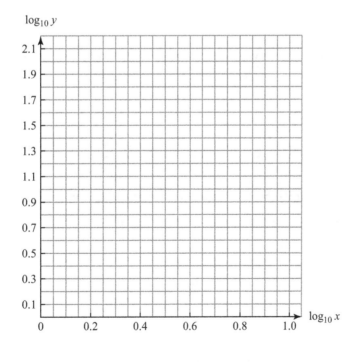

_____

_____

_____

_____

**7 b**    Estimate the values of the constants $k$ and $n$, giving your answers to
1 decimal place.       **[4 marks]**

**8**      The graph shows the gradient function, $y = f'(x)$, of a function $y = f(x)$

**a**      On the same axes, sketch the graph of $y = f''(x)$       **[3 marks]**

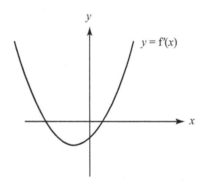

**b**      On this copy of the graph of $y = f'(x)$, sketch the graph of the original function $y = f(x)$

You may assume that $y = f(x)$ passes through the origin.       **[3 marks]**

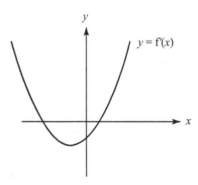

**9**  **a**    Three points, A, B and C, have position vectors $(2\mathbf{i} + 3\mathbf{j})$, $(6\mathbf{i} - 4\mathbf{j})$ and $(-5\mathbf{i} - 3\mathbf{j})$ respectively.

What is the perimeter of the triangle ABC?

Give your answer to 3 significant figures.                              **[5 marks]**

**b**    In the diagram, $\overrightarrow{OD} = \mathbf{d}$, $\overrightarrow{OE} = \mathbf{e}$ and $\overrightarrow{OF} = \mathbf{f}$

The midpoints of $\overrightarrow{EF}, \overrightarrow{FD}$ and $\overrightarrow{DE}$ are G, H and J respectively.

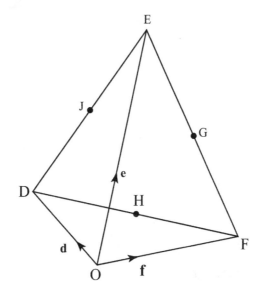

**i**    Find, in terms of **d**, **e** and **f**, the sum $\overrightarrow{OG} + \overrightarrow{OH} + \overrightarrow{OJ}$                              **[5 marks]**

**9 b ii** Thus write down $\overrightarrow{OG}+\overrightarrow{OH}+\overrightarrow{OJ}$ in terms of $\overrightarrow{OD}$, $\overrightarrow{OE}$ and $\overrightarrow{OF}$ [1 mark]

**10 a** On the same set of axes, given below, sketch the graphs of

**i** $y=\log_{10}x$ and [2 marks]

**ii** $y=10^x$ [2 marks]

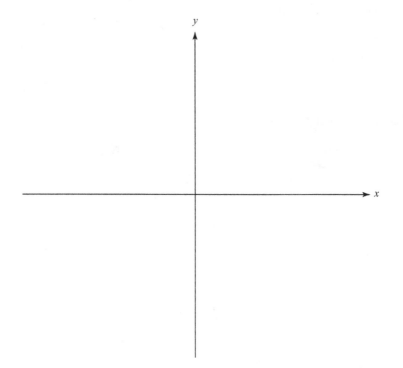

**iii** What is the relationship between $y=\log_{10}x$ and $y=10^x$?

How is this relationship shown in their graphs? [2 marks]

**10 b**   Solve this equation.   $10^{2x+5} = 4$

Give your answer to 3 significant figures.   **[3 marks]**

**11 a**   Three points A, B and C have coordinates $(-4, 6)$, $(4, 4)$ and $(1, -2)$ respectively.

**i**   Find the equation of the perpendicular bisector of the line AB   **[4 marks]**

**ii**   Find the equation of the perpendicular bisector of the line BC   **[1 mark]**

**11 b** Calculate the coordinates of the point of intersection of the perpendicular
bisectors to the lines AB and BC                                           [4 marks]

**c** Hence find the equation of the circumcircle of the triangle with vertices A, B and C.

Write your answer in the form $(x - a)^2 + (y - b)^2 = r^2$

where $a$, $b$ and $r^2$ are rational numbers.                             [4 marks]

**12 a** Complete this binomial expansion which has been started for you.

$$(a+b)^8 = a^8 + 8a^7b + 28a^6b^2 + ...$$

[2 marks]

**b** A multiple choice test consists of eight questions.

Each question has three answers to choose from, only one of which is correct.

Sidney tries to guess the answer to each of the eight questions.

What is the probability, to 3 significant figures, that Sidney guesses none of the answers correctly?

[2 marks]

**End of questions**

# AS Level Mathematics
## *Paper 2 (Set C)*

# Edexcel

| Name | | Class | |
|---|---|---|---|
| Signature | | Date | |

## Materials

Any calculator permitted by ICE regulations may be used. Calculators must not have retrievable mathematical formulae stored in them and they must not be able to differentiate, integrate or manipulate algebra.

## Instructions

- Use a black or blue pen.
- Use a dark pencil if not using a pen, for diagrams/graphs.
- Write your name and class at the top of this page in the box provided.
- Answer all the questions in this paper.
- Answer the questions in the spaces given.
- You may not need to use all the space given.
- You should make your methods clear by showing sufficient working or you may not gain full credit for your answers.
- Give inexact answers to three significant figures, unless otherwise stated.

## Information

- Extracts from the booklet 'Mathematical Formulae and Statistical Tables' have been provided.
- This question paper has 9 questions.
- There are 60 marks in total for this paper.
- The marks available for each question are in brackets.
- You should aim to spend approximately one minute per mark on each question.

## Advice

- Read each question carefully before you begin your answer.
- Attempt to answer all the questions.
- If you have time at the end, check your answers.
- If you start an answer again then cross out the old workings.

| Question | Mark |
|---|---|
| 1 | |
| 2 | |
| 3 | |
| 4 | |
| 5 | |
| 6 | |
| 7 | |
| 8 | |
| 9 | |
| Total | |

# Section A

Answer **all** questions in the spaces provided.

**1** A bag contains a large number of objects of varying colours and shapes.
These are listed in the table.

| | Blue | Green | Red | Total |
|---|---|---|---|---|
| **Circle** | 13 | 2 | 29 | 44 |
| **Square** | 16 | 6 | 6 | 28 |
| **Triangle** | 19 | 7 | 14 | 40 |
| **Total** | 48 | 15 | 49 | 112 |

**a** Calculate the probability that a randomly-chosen shape from the bag

**i** Is a blue square, **[2 marks]**

**ii** Is a square given that it is blue. **[2 marks]**

**b** Determine if the shape being red and the shape being a triangle are independent properties. Explain your reasoning. **[2 marks]**

**1    c**    Draw a Venn diagram to show the probabilities of the events of being red or not red
and of being a triangle or not a triangle.                                      **[2 marks]**

**2**    For a binomial distribution $X \sim \text{Bin}(18, 0.4)$ calculate the following probabilities.

**a**    $P(X = 3)$                                                                **[1 mark]**

**b**    $P(X \leq 5)$                                                              **[1 mark]**

**c**    $P(X > 8)$                                                                 **[1 mark]**

**d**    A survey finds that 5 out of 18 people prefer product A to product B.

The previous year, the survey showed that 40% of people preferred product A to product B.

At the 20% significance level, has the preference for product A decreased?

Explain your reasoning.                                                           **[3 marks]**

**3**  A manufacturer wants to investigate the preferred percentage of cocoa powder to use in its chocolate bars. It decides to survey 120 people in its local town. It believes that different groups of people will have different preferences so makes sure that its sample of people has the proportions of each group specified in the table below.

| Appearance | Office worker | Shop worker | Student | Retired | Other |
|---|---|---|---|---|---|
| % of population | 25% | 25% | 10% | 15% | 25% |
| Number in sample | 30 | 30 | 12 | 18 | 30 |

**a**  What is the name of this type of sampling method?                          **[1 mark]**

_____

_____

_____

**b**  The following data are taken from the 12 students sampled showing their preferred percentage of cocoa in chocolate bars. Higher percentages are for darker chocolate.

1%   5%   9%   21%   22%   25%   31%   37%   38%   60%   63%   68%

Calculate the five-number summary and interquartile range for the data.          **[4 marks]**

_____

_____

_____

_____

_____

_____

_____

_____

_____

_____

_____

_____

_____

_____

_____

_____

_____

_____

_____

_____

**3** **c**    The data for the office workers are collected and given in this box plot.

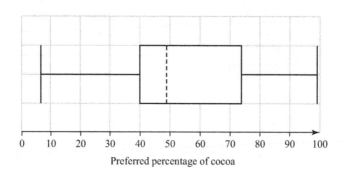

Preferred percentage of cocoa

Give three different comparisons between the preferred percentage of cocoa of the
students and the office workers.                                                   **[3 marks]**

**4**  People in a countryside town and in a coastal town are sampled to see if there are differences between the ages of the residents.

The results of the survey are presented in this cumulative frequency graph.

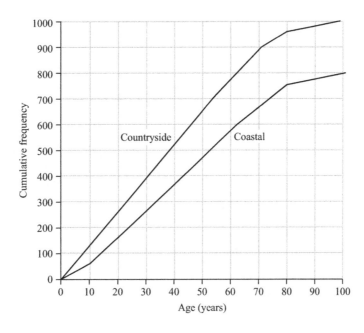

**a**  Estimate the percentage of people in each town who are 65 years or older.  **[3 marks]**

_____

_____

_____

_____

_____

_____

_____

_____

**b**  The mayor of the coastal town claims that the average age of someone in their town is 43 years. Explain how they determined that number.  **[2 marks]**

_____

_____

_____

_____

_____

_____

_____

_____

**4    c**    It is believed that political preference is associated with age.

Explain why cluster sampling by town should not be used to survey political preferences.

State which assumption in the use of cluster sampling is not present.                    **[3 marks]**

**End of section A**

# Section B

Answer **all** questions in the spaces provided.

**5**    A canoeist can paddle in still water at 4 km h$^{-1}$

She wants to paddle straight across a river to the opposite bank.

The water is flowing with a current of 3 km h$^{-1}$

**a**    At what angle to the riverbank must she paddle?    **[2 mark]**

**b**    The river is 22 m wide. How long will it take the canoeist to paddle straight across?    **[2 marks]**

**6**    A particle of mass 3 kg is acted on by forces of 5**i** N and 7**j** N.

Calculate the acceleration of the particle.    **[3 marks]**

**7  a**    When a skydiver jumps from an aircraft he accelerates until he reaches his *terminal velocity*.

The terminal velocity of a skydiver is 192 km h$^{-1}$

Convert 192 km h$^{-1}$ into m s$^{-1}$                                                   **[1 mark]**

_____

_____

_____

_____

_____

_____

_____

_____

_____

_____

**b**    Jo says that she should **not** use the formula $v = u + at$ to calculate the time it takes the skydiver to reach terminal velocity.

Explain why Jo is correct.                                                              **[1 mark]**

_____

_____

_____

_____

_____

_____

_____

_____

_____

_____

_____

_____

_____

_____

_____

_____

**7** **c** When a skydiver reaches terminal velocity he stops accelerating.

Describe the forces acting on a skydiver falling at terminal velocity and calculate the air resistance acting on a skydiver of mass 80 kg.

You should use $g = 9.81 \text{ m s}^{-2}$ [3 marks]

_____

_____

_____

_____

_____

_____

_____

**8** This velocity-time graph represents the motion of a particle moving along a straight line.

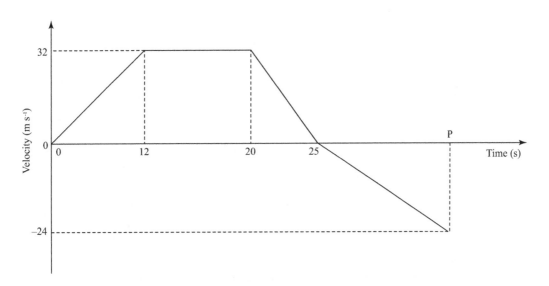

**a** Calculate the acceleration of the particle during the first 12 seconds. [2 marks]

_____

_____

_____

_____

_____

_____

_____

_____

_____

**8 b** When is the particle at rest? [1 mark]

**c** Given that the particle ends up in the same place that it started, find P. [5 marks]

**9** A particle starts from the origin with initial velocity $3 \, \text{m s}^{-1}$ and moves such that its acceleration after time $t$ s is given by $a = 6t - 4$

**a** Find a formula for the velocity, $v \, \text{m s}^{-1}$, of the particle after $t$ seconds. [3 marks]

**9   b**   Show that the particle never returns to the origin.                    **[7 marks]**

**End of questions**

# Mathematical formulae
## For AS Level Maths

## Pure Mathematics
### Mensuration

Surface area of sphere $= 4\pi r^2$

Area of curved surface of cone $= \pi r \times$ slant height

### Binomial series

$$(a+b)^n = a^n + \binom{n}{1}a^{n-1}b + \binom{n}{2}a^{n-2}b^2 + \ldots + \binom{n}{r}a^{n-r}b^r + \ldots + b^n \qquad (n \in \mathbb{N})$$

where $\binom{n}{r} = {}^nC_r = \dfrac{n!}{r!(n-r)!}$

### Logarithms and exponentials

$\log_a x = \dfrac{\log_b x}{\log_b a}$

$e^{x \ln a} = a^x$

### Differentiation
#### First principles

$$f'(x) = \lim_{h \to 0} \frac{f(x+h) - f(x)}{h}$$

## Statistics
### Probability

$P(A') = 1 - P(A)$

### Standard deviation

Standard deviation $= \sqrt{(\text{Variance})}$

Interquartile range $= \text{IQR} = Q_3 - Q_1$

For a set of $n$ values $x_1, x_2, \ldots, x_i \ldots x_n$

$$S_{xx} = \sum(x_i - \bar{x})^2 = \sum x_i^2 - \frac{\left(\sum x_i\right)^2}{n}$$

Standard deviation $= \sqrt{\dfrac{S_{xx}}{n}}$ or $\sqrt{\dfrac{\sum x^2}{n} - \bar{x}^2}$

### Discrete distributions

| Distribution of $X$ | $P(X = x)$ |
|---|---|
| Binomial $B(n, p)$ | $\binom{n}{x}p^x(1-p)^{n-x}$ |

## Mechanics
### Kinematics

For motion in a straight line with constant acceleration:

$v = u + at$

$s = ut + \dfrac{1}{2}at^2$

$s = vt - \dfrac{1}{2}at^2$

$v^2 = u^2 + 2as$

$s = \dfrac{1}{2}(u+v)t$

# Statistical tables
## For AS Level Maths

### Binomial Cumulative Distribution Function

The tabulated value is $P(X \leq x)$, where $X$ has a binomial distribution with index $n$ and parameter $p$

| $p =$ | 0.05 | 0.10 | 0.15 | 0.20 | 0.25 | 0.30 | 0.35 | 0.40 | 0.45 | 0.50 |
|---|---|---|---|---|---|---|---|---|---|---|
| $n = 5, x = 0$ | 0.7738 | 0.5905 | 0.4437 | 0.3277 | 0.2373 | 0.1681 | 0.1160 | 0.0778 | 0.0503 | 0.0312 |
| 1 | 0.9774 | 0.9185 | 0.8352 | 0.7373 | 0.6328 | 0.5282 | 0.4284 | 0.3370 | 0.2562 | 0.1875 |
| 2 | 0.9988 | 0.9914 | 0.9734 | 0.9421 | 0.8965 | 0.8369 | 0.7648 | 0.6826 | 0.5931 | 0.5000 |
| 3 | 1.0000 | 0.9995 | 0.9978 | 0.9933 | 0.9844 | 0.9692 | 0.9460 | 0.9130 | 0.8688 | 0.8125 |
| 4 | 1.0000 | 1.0000 | 0.9999 | 0.9997 | 0.9990 | 0.9976 | 0.9947 | 0.9898 | 0.9815 | 0.9688 |
| $n = 6, x = 0$ | 0.7351 | 0.5314 | 0.3771 | 0.2621 | 0.1780 | 0.1176 | 0.0754 | 0.0467 | 0.0277 | 0.0156 |
| 1 | 0.9672 | 0.8857 | 0.7765 | 0.6554 | 0.5339 | 0.4202 | 0.3191 | 0.2333 | 0.1636 | 0.1094 |
| 2 | 0.9978 | 0.9842 | 0.9527 | 0.9011 | 0.8306 | 0.7443 | 0.6471 | 0.5443 | 0.4415 | 0.3438 |
| 3 | 0.9999 | 0.9987 | 0.9941 | 0.9830 | 0.9624 | 0.9295 | 0.8826 | 0.8208 | 0.7447 | 0.6563 |
| 4 | 1.0000 | 0.9999 | 0.9996 | 0.9984 | 0.9954 | 0.9891 | 0.9777 | 0.9590 | 0.9308 | 0.8906 |
| 5 | 1.0000 | 1.0000 | 1.0000 | 0.9999 | 0.9998 | 0.9993 | 0.9982 | 0.9959 | 0.9917 | 0.9844 |
| $n = 7, x = 0$ | 0.6983 | 0.4783 | 0.3206 | 0.2097 | 0.1335 | 0.0824 | 0.0490 | 0.0280 | 0.0152 | 0.0078 |
| 1 | 0.9556 | 0.8503 | 0.7166 | 0.5767 | 0.4449 | 0.3294 | 0.2338 | 0.1586 | 0.1024 | 0.0625 |
| 2 | 0.9962 | 0.9743 | 0.9262 | 0.8520 | 0.7564 | 0.6471 | 0.5323 | 0.4199 | 0.3164 | 0.2266 |
| 3 | 0.9998 | 0.9973 | 0.9879 | 0.9667 | 0.9294 | 0.8740 | 0.8002 | 0.7102 | 0.6083 | 0.5000 |
| 4 | 1.0000 | 0.9998 | 0.9988 | 0.9953 | 0.9871 | 0.9712 | 0.9444 | 0.9037 | 0.8471 | 0.7734 |
| 5 | 1.0000 | 1.0000 | 0.9999 | 0.9996 | 0.9987 | 0.9962 | 0.9910 | 0.9812 | 0.9643 | 0.9375 |
| 6 | 1.0000 | 1.0000 | 1.0000 | 1.0000 | 0.9999 | 0.9998 | 0.9994 | 0.9984 | 0.9963 | 0.9922 |
| $n = 8, x = 0$ | 0.6634 | 0.4305 | 0.2725 | 0.1678 | 0.1001 | 0.0576 | 0.0319 | 0.0168 | 0.0084 | 0.0039 |
| 1 | 0.9428 | 0.8131 | 0.6572 | 0.5033 | 0.3671 | 0.2553 | 0.1691 | 0.1064 | 0.0632 | 0.0352 |
| 2 | 0.9942 | 0.9619 | 0.8948 | 0.7969 | 0.6785 | 0.5518 | 0.4278 | 0.3154 | 0.2201 | 0.1445 |
| 3 | 0.9996 | 0.9950 | 0.9786 | 0.9437 | 0.8862 | 0.8059 | 0.7064 | 0.5941 | 0.4770 | 0.3633 |
| 4 | 1.0000 | 0.9996 | 0.9971 | 0.9896 | 0.9727 | 0.9420 | 0.8939 | 0.8263 | 0.7396 | 0.6367 |
| 5 | 1.0000 | 1.0000 | 0.9998 | 0.9988 | 0.9958 | 0.9887 | 0.9747 | 0.9502 | 0.9115 | 0.8555 |
| 6 | 1.0000 | 1.0000 | 1.0000 | 0.9999 | 0.9996 | 0.9987 | 0.9964 | 0.9915 | 0.9819 | 0.9648 |
| 7 | 1.0000 | 1.0000 | 1.0000 | 1.0000 | 1.0000 | 0.9999 | 0.9998 | 0.9993 | 0.9983 | 0.9961 |
| $n = 9, x = 0$ | 0.6302 | 0.3874 | 0.2316 | 0.1342 | 0.0751 | 0.0404 | 0.0207 | 0.0101 | 0.0046 | 0.0020 |
| 1 | 0.9288 | 0.7748 | 0.5995 | 0.4362 | 0.3003 | 0.1960 | 0.1211 | 0.0705 | 0.0385 | 0.0195 |
| 2 | 0.9916 | 0.9470 | 0.8591 | 0.7382 | 0.6007 | 0.4628 | 0.3373 | 0.2318 | 0.1495 | 0.0898 |
| 3 | 0.9994 | 0.9917 | 0.9661 | 0.9144 | 0.8343 | 0.7297 | 0.6089 | 0.4826 | 0.3614 | 0.2539 |
| 4 | 1.0000 | 0.9991 | 0.9944 | 0.9804 | 0.9511 | 0.9012 | 0.8283 | 0.7334 | 0.6214 | 0.5000 |
| 5 | 1.0000 | 0.9999 | 0.9994 | 0.9969 | 0.9900 | 0.9747 | 0.9464 | 0.9006 | 0.8342 | 0.7461 |
| 6 | 1.0000 | 1.0000 | 1.0000 | 0.9997 | 0.9987 | 0.9957 | 0.9888 | 0.9750 | 0.9502 | 0.9102 |
| 7 | 1.0000 | 1.0000 | 1.0000 | 1.0000 | 0.9999 | 0.9996 | 0.9986 | 0.9962 | 0.9909 | 0.9805 |
| 8 | 1.0000 | 1.0000 | 1.0000 | 1.0000 | 1.0000 | 1.0000 | 0.9999 | 0.9997 | 0.9992 | 0.9980 |

| $p=$ | 0.05 | 0.10 | 0.15 | 0.20 | 0.25 | 0.30 | 0.35 | 0.40 | 0.45 | 0.50 |
|---|---|---|---|---|---|---|---|---|---|---|
| $n=10, x=0$ | 0.5987 | 0.3487 | 0.1969 | 0.1074 | 0.0563 | 0.0282 | 0.0135 | 0.0060 | 0.0025 | 0.0010 |
| 1 | 0.9139 | 0.7361 | 0.5443 | 0.3758 | 0.2440 | 0.1493 | 0.0860 | 0.0464 | 0.0233 | 0.0107 |
| 2 | 0.9885 | 0.9298 | 0.8202 | 0.6778 | 0.5256 | 0.3828 | 0.2616 | 0.1673 | 0.0996 | 0.0547 |
| 3 | 0.9990 | 0.9872 | 0.9500 | 0.8791 | 0.7759 | 0.6496 | 0.5138 | 0.3823 | 0.2660 | 0.1719 |
| 4 | 0.9999 | 0.9984 | 0.9901 | 0.9672 | 0.9219 | 0.8497 | 0.7515 | 0.6331 | 0.5044 | 0.3770 |
| 5 | 1.0000 | 0.9999 | 0.9986 | 0.9936 | 0.9803 | 0.9527 | 0.9051 | 0.8338 | 0.7384 | 0.6230 |
| 6 | 1.0000 | 1.0000 | 0.9999 | 0.9991 | 0.9965 | 0.9894 | 0.9740 | 0.9452 | 0.8980 | 0.8281 |
| 7 | 1.0000 | 1.0000 | 1.0000 | 0.9999 | 0.9996 | 0.9984 | 0.9952 | 0.9877 | 0.9726 | 0.9453 |
| 8 | 1.0000 | 1.0000 | 1.0000 | 1.0000 | 1.0000 | 0.9999 | 0.9995 | 0.9983 | 0.9955 | 0.9893 |
| 9 | 1.0000 | 1.0000 | 1.0000 | 1.0000 | 1.0000 | 1.0000 | 1.0000 | 0.9999 | 0.9997 | 0.9990 |
| $n=12, x=0$ | 0.5404 | 0.2824 | 0.1422 | 0.0687 | 0.0317 | 0.0138 | 0.0057 | 0.0022 | 0.0008 | 0.0002 |
| 1 | 0.8816 | 0.6590 | 0.4435 | 0.2749 | 0.1584 | 0.0850 | 0.0424 | 0.0196 | 0.0083 | 0.0032 |
| 2 | 0.9804 | 0.8891 | 0.7358 | 0.5583 | 0.3907 | 0.2528 | 0.1513 | 0.0834 | 0.0421 | 0.0193 |
| 3 | 0.9978 | 0.9744 | 0.9078 | 0.7946 | 0.6488 | 0.4925 | 0.3467 | 0.2253 | 0.1345 | 0.0730 |
| 4 | 0.9998 | 0.9957 | 0.9761 | 0.9274 | 0.8424 | 0.7237 | 0.5833 | 0.4382 | 0.3044 | 0.1938 |
| 5 | 1.0000 | 0.9995 | 0.9954 | 0.9806 | 0.9456 | 0.8822 | 0.7873 | 0.6652 | 0.5269 | 0.3872 |
| 6 | 1.0000 | 0.9999 | 0.9993 | 0.9961 | 0.9857 | 0.9614 | 0.9154 | 0.8418 | 0.7393 | 0.6128 |
| 7 | 1.0000 | 1.0000 | 0.9999 | 0.9994 | 0.9972 | 0.9905 | 0.9745 | 0.9427 | 0.8883 | 0.8062 |
| 8 | 1.0000 | 1.0000 | 1.0000 | 0.9999 | 0.9996 | 0.9983 | 0.9944 | 0.9847 | 0.9644 | 0.9270 |
| 9 | 1.0000 | 1.0000 | 1.0000 | 1.0000 | 1.0000 | 0.9998 | 0.9992 | 0.9972 | 0.9921 | 0.9807 |
| 10 | 1.0000 | 1.0000 | 1.0000 | 1.0000 | 1.0000 | 1.0000 | 0.9999 | 0.9997 | 0.9989 | 0.9968 |
| 11 | 1.0000 | 1.0000 | 1.0000 | 1.0000 | 1.0000 | 1.0000 | 1.0000 | 1.0000 | 0.9999 | 0.9998 |
| $n=15, x=0$ | 0.4633 | 0.2059 | 0.0874 | 0.0352 | 0.0134 | 0.0047 | 0.0016 | 0.0005 | 0.0001 | 0.0000 |
| 1 | 0.8290 | 0.5490 | 0.3186 | 0.1671 | 0.0802 | 0.0353 | 0.0142 | 0.0052 | 0.0017 | 0.0005 |
| 2 | 0.9638 | 0.8159 | 0.6042 | 0.3980 | 0.2361 | 0.1268 | 0.0617 | 0.0271 | 0.0107 | 0.0037 |
| 3 | 0.9945 | 0.9444 | 0.8227 | 0.6482 | 0.4613 | 0.2969 | 0.1727 | 0.0905 | 0.0424 | 0.0176 |
| 4 | 0.9994 | 0.9873 | 0.9383 | 0.8358 | 0.6865 | 0.5155 | 0.3519 | 0.2173 | 0.1204 | 0.0592 |
| 5 | 0.9999 | 0.9978 | 0.9832 | 0.9389 | 0.8516 | 0.7216 | 0.5643 | 0.4032 | 0.2608 | 0.1509 |
| 6 | 1.0000 | 0.9997 | 0.9964 | 0.9819 | 0.9434 | 0.8689 | 0.7548 | 0.6098 | 0.4522 | 0.3036 |
| 7 | 1.0000 | 1.0000 | 0.9994 | 0.9958 | 0.9827 | 0.9500 | 0.8868 | 0.7869 | 0.6535 | 0.5000 |
| 8 | 1.0000 | 1.0000 | 0.9999 | 0.9992 | 0.9958 | 0.9848 | 0.9578 | 0.9050 | 0.8182 | 0.6964 |
| 9 | 1.0000 | 1.0000 | 1.0000 | 0.9999 | 0.9992 | 0.9963 | 0.9876 | 0.9662 | 0.9231 | 0.8491 |
| 10 | 1.0000 | 1.0000 | 1.0000 | 1.0000 | 0.9999 | 0.9993 | 0.9972 | 0.9907 | 0.9745 | 0.9408 |
| 11 | 1.0000 | 1.0000 | 1.0000 | 1.0000 | 1.0000 | 0.9999 | 0.9995 | 0.9981 | 0.9937 | 0.9824 |
| 12 | 1.0000 | 1.0000 | 1.0000 | 1.0000 | 1.0000 | 1.0000 | 0.9999 | 0.9997 | 0.9989 | 0.9963 |
| 13 | 1.0000 | 1.0000 | 1.0000 | 1.0000 | 1.0000 | 1.0000 | 1.0000 | 1.0000 | 0.9999 | 0.9995 |
| 14 | 1.0000 | 1.0000 | 1.0000 | 1.0000 | 1.0000 | 1.0000 | 1.0000 | 1.0000 | 1.0000 | 1.0000 |

| p= | 0.05 | 0.10 | 0.15 | 0.20 | 0.25 | 0.30 | 0.35 | 0.40 | 0.45 | 0.50 |
|---|---|---|---|---|---|---|---|---|---|---|
| n = 20, x = 0 | 0.3585 | 0.1216 | 0.0388 | 0.0115 | 0.0032 | 0.0008 | 0.0002 | 0.0000 | 0.0000 | 0.0000 |
| 1 | 0.7358 | 0.3917 | 0.1756 | 0.0692 | 0.0243 | 0.0076 | 0.0021 | 0.0005 | 0.0001 | 0.0000 |
| 2 | 0.9245 | 0.6769 | 0.4049 | 0.2061 | 0.0913 | 0.0355 | 0.0121 | 0.0036 | 0.0009 | 0.0002 |
| 3 | 0.9841 | 0.8670 | 0.6477 | 0.4114 | 0.2252 | 0.1071 | 0.0444 | 0.0160 | 0.0049 | 0.0013 |
| 4 | 0.9974 | 0.9568 | 0.8298 | 0.6296 | 0.4148 | 0.2375 | 0.1182 | 0.0510 | 0.0189 | 0.0059 |
| 5 | 0.9997 | 0.9887 | 0.9327 | 0.8042 | 0.6172 | 0.4164 | 0.2454 | 0.1256 | 0.0553 | 0.0207 |
| 6 | 1.0000 | 0.9976 | 0.9781 | 0.9133 | 0.7858 | 0.6080 | 0.4166 | 0.2500 | 0.1299 | 0.0577 |
| 7 | 1.0000 | 0.9996 | 0.9941 | 0.9679 | 0.8982 | 0.7723 | 0.6010 | 0.4159 | 0.2520 | 0.1316 |
| 8 | 1.0000 | 0.9999 | 0.9987 | 0.9900 | 0.9591 | 0.8867 | 0.7624 | 0.5956 | 0.4143 | 0.2517 |
| 9 | 1.0000 | 1.0000 | 0.9998 | 0.9974 | 0.9861 | 0.9520 | 0.8782 | 0.7553 | 0.5914 | 0.4119 |
| 10 | 1.0000 | 1.0000 | 1.0000 | 0.9994 | 0.9961 | 0.9829 | 0.9468 | 0.8725 | 0.7507 | 0.5881 |
| 11 | 1.0000 | 1.0000 | 1.0000 | 0.9999 | 0.9991 | 0.9949 | 0.9804 | 0.9435 | 0.8692 | 0.7483 |
| 12 | 1.0000 | 1.0000 | 1.0000 | 1.0000 | 0.9998 | 0.9987 | 0.9940 | 0.9790 | 0.9420 | 0.8684 |
| 13 | 1.0000 | 1.0000 | 1.0000 | 1.0000 | 1.0000 | 0.9997 | 0.9985 | 0.9935 | 0.9786 | 0.9423 |
| 14 | 1.0000 | 1.0000 | 1.0000 | 1.0000 | 1.0000 | 1.0000 | 0.9997 | 0.9984 | 0.9936 | 0.9793 |
| 15 | 1.0000 | 1.0000 | 1.0000 | 1.0000 | 1.0000 | 1.0000 | 1.0000 | 0.9997 | 0.9985 | 0.9941 |
| 16 | 1.0000 | 1.0000 | 1.0000 | 1.0000 | 1.0000 | 1.0000 | 1.0000 | 1.0000 | 0.9997 | 0.9987 |
| 17 | 1.0000 | 1.0000 | 1.0000 | 1.0000 | 1.0000 | 1.0000 | 1.0000 | 1.0000 | 1.0000 | 0.9998 |
| 18 | 1.0000 | 1.0000 | 1.0000 | 1.0000 | 1.0000 | 1.0000 | 1.0000 | 1.0000 | 1.0000 | 1.0000 |
| n = 25, x = 0 | 0.2774 | 0.0718 | 0.0172 | 0.0038 | 0.0008 | 0.0001 | 0.0000 | 0.0000 | 0.0000 | 0.0000 |
| 1 | 0.6424 | 0.2712 | 0.0931 | 0.0274 | 0.0070 | 0.0016 | 0.0003 | 0.0001 | 0.0000 | 0.0000 |
| 2 | 0.8729 | 0.5371 | 0.2537 | 0.0982 | 0.0321 | 0.0090 | 0.0021 | 0.0004 | 0.0001 | 0.0000 |
| 3 | 0.9659 | 0.7636 | 0.4711 | 0.2340 | 0.0962 | 0.0332 | 0.0097 | 0.0024 | 0.0005 | 0.0001 |
| 4 | 0.9928 | 0.9020 | 0.6821 | 0.4207 | 0.2137 | 0.0905 | 0.0320 | 0.0095 | 0.0023 | 0.0005 |
| 5 | 0.9988 | 0.9666 | 0.8385 | 0.6167 | 0.3783 | 0.1935 | 0.0826 | 0.0294 | 0.0086 | 0.0020 |
| 6 | 0.9998 | 0.9905 | 0.9305 | 0.7800 | 0.5611 | 0.3407 | 0.1734 | 0.0736 | 0.0258 | 0.0073 |
| 7 | 1.0000 | 0.9977 | 0.9745 | 0.8909 | 0.7265 | 0.5118 | 0.3061 | 0.1536 | 0.0639 | 0.0216 |
| 8 | 1.0000 | 0.9995 | 0.9920 | 0.9532 | 0.8506 | 0.6769 | 0.4668 | 0.2735 | 0.1340 | 0.0539 |
| 9 | 1.0000 | 0.9999 | 0.9979 | 0.9827 | 0.9287 | 0.8106 | 0.6303 | 0.4246 | 0.2424 | 0.1148 |
| 10 | 1.0000 | 1.0000 | 0.9995 | 0.9944 | 0.9703 | 0.9022 | 0.7712 | 0.5858 | 0.3843 | 0.2122 |
| 11 | 1.0000 | 1.0000 | 0.9999 | 0.9985 | 0.9893 | 0.9558 | 0.8746 | 0.7323 | 0.5426 | 0.3450 |
| 12 | 1.0000 | 1.0000 | 1.0000 | 0.9996 | 0.9966 | 0.9825 | 0.9396 | 0.8462 | 0.6937 | 0.5000 |
| 13 | 1.0000 | 1.0000 | 1.0000 | 0.9999 | 0.9991 | 0.9940 | 0.9745 | 0.9222 | 0.8173 | 0.6550 |
| 14 | 1.0000 | 1.0000 | 1.0000 | 1.0000 | 0.9998 | 0.9982 | 0.9907 | 0.9656 | 0.9040 | 0.7878 |
| 15 | 1.0000 | 1.0000 | 1.0000 | 1.0000 | 1.0000 | 0.9995 | 0.9971 | 0.9868 | 0.9560 | 0.8852 |
| 16 | 1.0000 | 1.0000 | 1.0000 | 1.0000 | 1.0000 | 0.9999 | 0.9992 | 0.9957 | 0.9826 | 0.9461 |
| 17 | 1.0000 | 1.0000 | 1.0000 | 1.0000 | 1.0000 | 1.0000 | 0.9998 | 0.9988 | 0.9942 | 0.9784 |
| 18 | 1.0000 | 1.0000 | 1.0000 | 1.0000 | 1.0000 | 1.0000 | 1.0000 | 0.9997 | 0.9984 | 0.9927 |
| 19 | 1.0000 | 1.0000 | 1.0000 | 1.0000 | 1.0000 | 1.0000 | 1.0000 | 0.9999 | 0.9996 | 0.9980 |
| 20 | 1.0000 | 1.0000 | 1.0000 | 1.0000 | 1.0000 | 1.0000 | 1.0000 | 1.0000 | 0.9999 | 0.9995 |
| 21 | 1.0000 | 1.0000 | 1.0000 | 1.0000 | 1.0000 | 1.0000 | 1.0000 | 1.0000 | 1.0000 | 0.9999 |
| 22 | 1.0000 | 1.0000 | 1.0000 | 1.0000 | 1.0000 | 1.0000 | 1.0000 | 1.0000 | 1.0000 | 1.0000 |

| $p=$ | 0.05 | 0.10 | 0.15 | 0.20 | 0.25 | 0.30 | 0.35 | 0.40 | 0.45 | 0.50 |
|---|---|---|---|---|---|---|---|---|---|---|
| $n=30, x=0$ | 0.2146 | 0.0424 | 0.0076 | 0.0012 | 0.0002 | 0.0000 | 0.0000 | 0.0000 | 0.0000 | 0.0000 |
| 1 | 0.5535 | 0.1837 | 0.0480 | 0.0105 | 0.0020 | 0.0003 | 0.0000 | 0.0000 | 0.0000 | 0.0000 |
| 2 | 0.8122 | 0.4114 | 0.1514 | 0.0442 | 0.0106 | 0.0021 | 0.0003 | 0.0000 | 0.0000 | 0.0000 |
| 3 | 0.9392 | 0.6474 | 0.3217 | 0.1227 | 0.0374 | 0.0093 | 0.0019 | 0.0003 | 0.0000 | 0.0000 |
| 4 | 0.9844 | 0.8245 | 0.5245 | 0.2552 | 0.0979 | 0.0302 | 0.0075 | 0.0015 | 0.0002 | 0.0000 |
| 5 | 0.9967 | 0.9268 | 0.7106 | 0.4275 | 0.2026 | 0.0766 | 0.0233 | 0.0057 | 0.0011 | 0.0002 |
| 6 | 0.9994 | 0.9742 | 0.8474 | 0.6070 | 0.3481 | 0.1595 | 0.0586 | 0.0172 | 0.0040 | 0.0007 |
| 7 | 0.9999 | 0.9922 | 0.9302 | 0.7608 | 0.5143 | 0.2814 | 0.1238 | 0.0435 | 0.0121 | 0.0026 |
| 8 | 1.0000 | 0.9980 | 0.9722 | 0.8713 | 0.6736 | 0.4315 | 0.2247 | 0.0940 | 0.0312 | 0.0081 |
| 9 | 1.0000 | 0.9995 | 0.9903 | 0.9389 | 0.8034 | 0.5888 | 0.3575 | 0.1763 | 0.0694 | 0.0214 |
| 10 | 1.0000 | 0.9999 | 0.9971 | 0.9744 | 0.8943 | 0.7304 | 0.5078 | 0.2915 | 0.1350 | 0.0494 |
| 11 | 1.0000 | 1.0000 | 0.9992 | 0.9905 | 0.9493 | 0.8407 | 0.6548 | 0.4311 | 0.2327 | 0.1002 |
| 12 | 1.0000 | 1.0000 | 0.9998 | 0.9969 | 0.9784 | 0.9155 | 0.7802 | 0.5785 | 0.3592 | 0.1808 |
| 13 | 1.0000 | 1.0000 | 1.0000 | 0.9991 | 0.9918 | 0.9599 | 0.8737 | 0.7145 | 0.5025 | 0.2923 |
| 14 | 1.0000 | 1.0000 | 1.0000 | 0.9998 | 0.9973 | 0.9831 | 0.9348 | 0.8246 | 0.6448 | 0.4278 |
| 15 | 1.0000 | 1.0000 | 1.0000 | 0.9999 | 0.9992 | 0.9936 | 0.9699 | 0.9029 | 0.7691 | 0.5722 |
| 16 | 1.0000 | 1.0000 | 1.0000 | 1.0000 | 0.9998 | 0.9979 | 0.9876 | 0.9519 | 0.8644 | 0.7077 |
| 17 | 1.0000 | 1.0000 | 1.0000 | 1.0000 | 0.9999 | 0.9994 | 0.9955 | 0.9788 | 0.9286 | 0.8192 |
| 18 | 1.0000 | 1.0000 | 1.0000 | 1.0000 | 1.0000 | 0.9998 | 0.9986 | 0.9917 | 0.9666 | 0.8998 |
| 19 | 1.0000 | 1.0000 | 1.0000 | 1.0000 | 1.0000 | 1.0000 | 0.9996 | 0.9971 | 0.9862 | 0.9506 |
| 20 | 1.0000 | 1.0000 | 1.0000 | 1.0000 | 1.0000 | 1.0000 | 0.9999 | 0.9991 | 0.9950 | 0.9786 |
| 21 | 1.0000 | 1.0000 | 1.0000 | 1.0000 | 1.0000 | 1.0000 | 1.0000 | 0.9998 | 0.9984 | 0.9919 |
| 22 | 1.0000 | 1.0000 | 1.0000 | 1.0000 | 1.0000 | 1.0000 | 1.0000 | 1.0000 | 0.9996 | 0.9974 |
| 23 | 1.0000 | 1.0000 | 1.0000 | 1.0000 | 1.0000 | 1.0000 | 1.0000 | 1.0000 | 0.9999 | 0.9993 |
| 24 | 1.0000 | 1.0000 | 1.0000 | 1.0000 | 1.0000 | 1.0000 | 1.0000 | 1.0000 | 1.0000 | 0.9998 |
| 25 | 1.0000 | 1.0000 | 1.0000 | 1.0000 | 1.0000 | 1.0000 | 1.0000 | 1.0000 | 1.0000 | 1.0000 |

| $p=$ | 0.05 | 0.10 | 0.15 | 0.20 | 0.25 | 0.30 | 0.35 | 0.40 | 0.45 | 0.50 |
|---|---|---|---|---|---|---|---|---|---|---|
| $n=40, x=0$ | 0.1285 | 0.0148 | 0.0015 | 0.0001 | 0.0000 | 0.0000 | 0.0000 | 0.0000 | 0.0000 | 0.0000 |
| 1 | 0.3991 | 0.0805 | 0.0121 | 0.0015 | 0.0001 | 0.0000 | 0.0000 | 0.0000 | 0.0000 | 0.0000 |
| 2 | 0.6767 | 0.2228 | 0.0486 | 0.0079 | 0.0010 | 0.0001 | 0.0000 | 0.0000 | 0.0000 | 0.0000 |
| 3 | 0.8619 | 0.4231 | 0.1302 | 0.0285 | 0.0047 | 0.0006 | 0.0001 | 0.0000 | 0.0000 | 0.0000 |
| 4 | 0.9520 | 0.6290 | 0.2633 | 0.0759 | 0.0160 | 0.0026 | 0.0003 | 0.0000 | 0.0000 | 0.0000 |
| 5 | 0.9861 | 0.7937 | 0.4325 | 0.1613 | 0.0433 | 0.0086 | 0.0013 | 0.0001 | 0.0000 | 0.0000 |
| 6 | 0.9966 | 0.9005 | 0.6067 | 0.2859 | 0.0962 | 0.0238 | 0.0044 | 0.0006 | 0.0001 | 0.0000 |
| 7 | 0.9993 | 0.9581 | 0.7559 | 0.4371 | 0.1820 | 0.0553 | 0.0124 | 0.0021 | 0.0002 | 0.0000 |
| 8 | 0.9999 | 0.9845 | 0.8646 | 0.5931 | 0.2998 | 0.1110 | 0.0303 | 0.0061 | 0.0009 | 0.0001 |
| 9 | 1.0000 | 0.9949 | 0.9328 | 0.7318 | 0.4395 | 0.1959 | 0.0644 | 0.0156 | 0.0027 | 0.0003 |
| 10 | 1.0000 | 0.9985 | 0.9701 | 0.8392 | 0.5839 | 0.3087 | 0.1215 | 0.0352 | 0.0074 | 0.0011 |
| 11 | 1.0000 | 0.9996 | 0.9880 | 0.9125 | 0.7151 | 0.4406 | 0.2053 | 0.0709 | 0.0179 | 0.0032 |
| 12 | 1.0000 | 0.9999 | 0.9957 | 0.9568 | 0.8209 | 0.5772 | 0.3143 | 0.1285 | 0.0386 | 0.0083 |
| 13 | 1.0000 | 1.0000 | 0.9986 | 0.9806 | 0.8968 | 0.7032 | 0.4408 | 0.2112 | 0.0751 | 0.0192 |
| 14 | 1.0000 | 1.0000 | 0.9996 | 0.9921 | 0.9456 | 0.8074 | 0.5721 | 0.3174 | 0.1326 | 0.0403 |
| 15 | 1.0000 | 1.0000 | 0.9999 | 0.9971 | 0.9738 | 0.8849 | 0.6946 | 0.4402 | 0.2142 | 0.0769 |
| 16 | 1.0000 | 1.0000 | 1.0000 | 0.9990 | 0.9884 | 0.9367 | 0.7978 | 0.5681 | 0.3185 | 0.1341 |
| 17 | 1.0000 | 1.0000 | 1.0000 | 0.9997 | 0.9953 | 0.9680 | 0.8761 | 0.6885 | 0.4391 | 0.2148 |
| 18 | 1.0000 | 1.0000 | 1.0000 | 0.9999 | 0.9983 | 0.9852 | 0.9301 | 0.7911 | 0.5651 | 0.3179 |
| 19 | 1.0000 | 1.0000 | 1.0000 | 1.0000 | 0.9994 | 0.9937 | 0.9637 | 0.8702 | 0.6844 | 0.4373 |
| 20 | 1.0000 | 1.0000 | 1.0000 | 1.0000 | 0.9998 | 0.9976 | 0.9827 | 0.9256 | 0.7870 | 0.5627 |
| 21 | 1.0000 | 1.0000 | 1.0000 | 1.0000 | 1.0000 | 0.9991 | 0.9925 | 0.9608 | 0.8669 | 0.6821 |
| 22 | 1.0000 | 1.0000 | 1.0000 | 1.0000 | 1.0000 | 0.9997 | 0.9970 | 0.9811 | 0.9233 | 0.7852 |
| 23 | 1.0000 | 1.0000 | 1.0000 | 1.0000 | 1.0000 | 0.9999 | 0.9989 | 0.9917 | 0.9595 | 0.8659 |
| 24 | 1.0000 | 1.0000 | 1.0000 | 1.0000 | 1.0000 | 1.0000 | 0.9996 | 0.9966 | 0.9804 | 0.9231 |
| 25 | 1.0000 | 1.0000 | 1.0000 | 1.0000 | 1.0000 | 1.0000 | 0.9999 | 0.9988 | 0.9914 | 0.9597 |
| 26 | 1.0000 | 1.0000 | 1.0000 | 1.0000 | 1.0000 | 1.0000 | 1.0000 | 0.9996 | 0.9966 | 0.9808 |
| 27 | 1.0000 | 1.0000 | 1.0000 | 1.0000 | 1.0000 | 1.0000 | 1.0000 | 0.9999 | 0.9988 | 0.9917 |
| 28 | 1.0000 | 1.0000 | 1.0000 | 1.0000 | 1.0000 | 1.0000 | 1.0000 | 1.0000 | 0.9996 | 0.9968 |
| 29 | 1.0000 | 1.0000 | 1.0000 | 1.0000 | 1.0000 | 1.0000 | 1.0000 | 1.0000 | 0.9999 | 0.9989 |
| 30 | 1.0000 | 1.0000 | 1.0000 | 1.0000 | 1.0000 | 1.0000 | 1.0000 | 1.0000 | 1.0000 | 0.9997 |
| 31 | 1.0000 | 1.0000 | 1.0000 | 1.0000 | 1.0000 | 1.0000 | 1.0000 | 1.0000 | 1.0000 | 0.9999 |
| 32 | 1.0000 | 1.0000 | 1.0000 | 1.0000 | 1.0000 | 1.0000 | 1.0000 | 1.0000 | 1.0000 | 1.0000 |

| $p =$ | 0.05 | 0.10 | 0.15 | 0.20 | 0.25 | 0.30 | 0.35 | 0.40 | 0.45 | 0.50 |
|---|---|---|---|---|---|---|---|---|---|---|
| $n = 50, x = 0$ | 0.0769 | 0.0052 | 0.0003 | 0.0000 | 0.0000 | 0.0000 | 0.0000 | 0.0000 | 0.0000 | 0.0000 |
| 1 | 0.2794 | 0.0338 | 0.0029 | 0.0002 | 0.0000 | 0.0000 | 0.0000 | 0.0000 | 0.0000 | 0.0000 |
| 2 | 0.5405 | 0.1117 | 0.0142 | 0.0013 | 0.0001 | 0.0000 | 0.0000 | 0.0000 | 0.0000 | 0.0000 |
| 3 | 0.7604 | 0.2503 | 0.0460 | 0.0057 | 0.0005 | 0.0000 | 0.0000 | 0.0000 | 0.0000 | 0.0000 |
| 4 | 0.8964 | 0.4312 | 0.1121 | 0.0185 | 0.0021 | 0.0002 | 0.0000 | 0.0000 | 0.0000 | 0.0000 |
| 5 | 0.9622 | 0.6161 | 0.2194 | 0.0480 | 0.0070 | 0.0007 | 0.0001 | 0.0000 | 0.0000 | 0.0000 |
| 6 | 0.9882 | 0.7702 | 0.3613 | 0.1034 | 0.0194 | 0.0025 | 0.0002 | 0.0000 | 0.0000 | 0.0000 |
| 7 | 0.9968 | 0.8779 | 0.5188 | 0.1904 | 0.0453 | 0.0073 | 0.0008 | 0.0001 | 0.0000 | 0.0000 |
| 8 | 0.9992 | 0.9421 | 0.6681 | 0.3073 | 0.0916 | 0.0183 | 0.0025 | 0.0002 | 0.0000 | 0.0000 |
| 9 | 0.9998 | 0.9755 | 0.7911 | 0.4437 | 0.1637 | 0.0402 | 0.0067 | 0.0008 | 0.0001 | 0.0000 |
| 10 | 1.0000 | 0.9906 | 0.8801 | 0.5836 | 0.2622 | 0.0789 | 0.0160 | 0.0022 | 0.0002 | 0.0000 |
| 11 | 1.0000 | 0.9968 | 0.9372 | 0.7107 | 0.3816 | 0.1390 | 0.0342 | 0.0057 | 0.0006 | 0.0000 |
| 12 | 1.0000 | 0.9990 | 0.9699 | 0.8139 | 0.5110 | 0.2229 | 0.0661 | 0.0133 | 0.0018 | 0.0002 |
| 13 | 1.0000 | 0.9997 | 0.9868 | 0.8894 | 0.6370 | 0.3279 | 0.1163 | 0.0280 | 0.0045 | 0.0005 |
| 14 | 1.0000 | 0.9999 | 0.9947 | 0.9393 | 0.7481 | 0.4468 | 0.1878 | 0.0540 | 0.0104 | 0.0013 |
| 15 | 1.0000 | 1.0000 | 0.9981 | 0.9692 | 0.8369 | 0.5692 | 0.2801 | 0.0955 | 0.0220 | 0.0033 |
| 16 | 1.0000 | 1.0000 | 0.9993 | 0.9856 | 0.9017 | 0.6839 | 0.3889 | 0.1561 | 0.0427 | 0.0077 |
| 17 | 1.0000 | 1.0000 | 0.9998 | 0.9937 | 0.9449 | 0.7822 | 0.5060 | 0.2369 | 0.0765 | 0.0164 |
| 18 | 1.0000 | 1.0000 | 0.9999 | 0.9975 | 0.9713 | 0.8594 | 0.6216 | 0.3356 | 0.1273 | 0.0325 |
| 19 | 1.0000 | 1.0000 | 1.0000 | 0.9991 | 0.9861 | 0.9152 | 0.7264 | 0.4465 | 0.1974 | 0.0595 |
| 20 | 1.0000 | 1.0000 | 1.0000 | 0.9997 | 0.9937 | 0.9522 | 0.8139 | 0.5610 | 0.2862 | 0.1013 |
| 21 | 1.0000 | 1.0000 | 1.0000 | 0.9999 | 0.9974 | 0.9749 | 0.8813 | 0.6701 | 0.3900 | 0.1611 |
| 22 | 1.0000 | 1.0000 | 1.0000 | 1.0000 | 0.9990 | 0.9877 | 0.9290 | 0.7660 | 0.5019 | 0.2399 |
| 23 | 1.0000 | 1.0000 | 1.0000 | 1.0000 | 0.9996 | 0.9944 | 0.9604 | 0.8438 | 0.6134 | 0.3359 |
| 24 | 1.0000 | 1.0000 | 1.0000 | 1.0000 | 0.9999 | 0.9976 | 0.9793 | 0.9022 | 0.7160 | 0.4439 |
| 25 | 1.0000 | 1.0000 | 1.0000 | 1.0000 | 1.0000 | 0.9991 | 0.9900 | 0.9427 | 0.8034 | 0.5561 |
| 26 | 1.0000 | 1.0000 | 1.0000 | 1.0000 | 1.0000 | 0.9997 | 0.9955 | 0.9686 | 0.8721 | 0.6641 |
| 27 | 1.0000 | 1.0000 | 1.0000 | 1.0000 | 1.0000 | 0.9999 | 0.9981 | 0.9840 | 0.9220 | 0.7601 |
| 28 | 1.0000 | 1.0000 | 1.0000 | 1.0000 | 1.0000 | 1.0000 | 0.9993 | 0.9924 | 0.9556 | 0.8389 |
| 29 | 1.0000 | 1.0000 | 1.0000 | 1.0000 | 1.0000 | 1.0000 | 0.9997 | 0.9966 | 0.9765 | 0.8987 |
| 30 | 1.0000 | 1.0000 | 1.0000 | 1.0000 | 1.0000 | 1.0000 | 0.9999 | 0.9986 | 0.9884 | 0.9405 |
| 31 | 1.0000 | 1.0000 | 1.0000 | 1.0000 | 1.0000 | 1.0000 | 1.0000 | 0.9995 | 0.9947 | 0.9675 |
| 32 | 1.0000 | 1.0000 | 1.0000 | 1.0000 | 1.0000 | 1.0000 | 1.0000 | 0.9998 | 0.9978 | 0.9836 |
| 33 | 1.0000 | 1.0000 | 1.0000 | 1.0000 | 1.0000 | 1.0000 | 1.0000 | 0.9999 | 0.9991 | 0.9923 |
| 34 | 1.0000 | 1.0000 | 1.0000 | 1.0000 | 1.0000 | 1.0000 | 1.0000 | 1.0000 | 0.9997 | 0.9967 |
| 35 | 1.0000 | 1.0000 | 1.0000 | 1.0000 | 1.0000 | 1.0000 | 1.0000 | 1.0000 | 0.9999 | 0.9987 |
| 36 | 1.0000 | 1.0000 | 1.0000 | 1.0000 | 1.0000 | 1.0000 | 1.0000 | 1.0000 | 1.0000 | 0.9995 |
| 37 | 1.0000 | 1.0000 | 1.0000 | 1.0000 | 1.0000 | 1.0000 | 1.0000 | 1.0000 | 1.0000 | 0.9998 |
| 38 | 1.0000 | 1.0000 | 1.0000 | 1.0000 | 1.0000 | 1.0000 | 1.0000 | 1.0000 | 1.0000 | 1.0000 |

## Random numbers

| | | | | | | | | | |
|---|---|---|---|---|---|---|---|---|---|
| 86 13 | 84 10 | 07 30 | 39 05 | 97 96 | 88 07 | 37 26 | 04 89 | 13 48 | 19 20 |
| 60 78 | 48 12 | 99 47 | 09 46 | 91 33 | 17 21 | 03 94 | 79 00 | 08 50 | 40 16 |
| 78 48 | 06 37 | 82 26 | 01 06 | 64 65 | 94 41 | 17 26 | 74 66 | 61 93 | 24 97 |
| 80 56 | 90 79 | 66 94 | 18 40 | 97 79 | 93 20 | 41 51 | 25 04 | 20 71 | 76 04 |
| 99 09 | 39 25 | 66 31 | 70 56 | 30 15 | 52 17 | 87 55 | 31 11 | 10 68 | 98 23 |
| 56 32 | 32 72 | 91 65 | 97 36 | 56 61 | 12 79 | 95 17 | 57 16 | 53 58 | 96 36 |
| 66 02 | 49 93 | 97 44 | 99 15 | 56 86 | 80 57 | 11 78 | 40 23 | 58 40 | 86 14 |
| 31 77 | 53 94 | 05 93 | 56 14 | 71 23 | 60 46 | 05 33 | 23 72 | 93 10 | 81 23 |
| 98 79 | 72 43 | 14 76 | 54 77 | 66 29 | 84 09 | 88 56 | 75 86 | 41 67 | 04 42 |
| 50 97 | 92 15 | 10 01 | 57 01 | 87 33 | 73 17 | 70 18 | 40 21 | 24 20 | 66 62 |
| 90 51 | 94 50 | 12 48 | 88 95 | 09 34 | 09 30 | 22 27 | 25 56 | 40 76 | 01 59 |
| 31 99 | 52 24 | 13 43 | 27 88 | 11 39 | 41 65 | 00 84 | 13 06 | 31 79 | 74 97 |
| 22 96 | 23 34 | 46 12 | 67 11 | 48 06 | 99 24 | 14 83 | 78 37 | 65 73 | 39 47 |
| 06 84 | 55 41 | 27 06 | 74 59 | 14 29 | 20 14 | 45 75 | 31 16 | 05 41 | 22 96 |
| 08 64 | 89 30 | 25 25 | 71 35 | 33 31 | 04 56 | 12 67 | 03 74 | 07 16 | 49 32 |
| 86 87 | 62 43 | 15 11 | 76 49 | 79 13 | 78 80 | 93 89 | 09 57 | 07 14 | 40 74 |
| 94 44 | 97 13 | 77 04 | 35 02 | 12 76 | 60 91 | 93 40 | 81 06 | 85 85 | 72 84 |
| 63 25 | 55 14 | 66 47 | 99 90 | 02 90 | 83 43 | 16 01 | 19 69 | 11 78 | 87 84 |
| 11 22 | 83 98 | 15 21 | 18 57 | 53 42 | 91 91 | 26 52 | 89 13 | 86 00 | 47 61 |
| 01 70 | 10 83 | 94 71 | 13 67 | 11 12 | 36 54 | 53 32 | 90 43 | 79 01 | 95 15 |

# Objectives checklist

Use these checklists to assess your confidence with the topics in AS Level Maths.

| Ch | Objective | MyMaths | InvisiPen | No | Almost | Yes! |
|---|---|---|---|---|---|---|
| **1 Algebra 1** | Use direct proof, proof by exhaustion and counter-examples. | 2252, 2253 | 01S1A | ☐ | ☐ | ☐ |
| | Use and manipulate index laws. | 2033–2035 | 01S2B | ☐ | ☐ | ☐ |
| | Manipulate surds and rationalise a denominator. | 2036, 2037 | 01S3A | ☐ | ☐ | ☐ |
| | Solve quadratic equations and sketch quadratic curves. | 2014–2017, 2024–2026 | 01S4B | ☐ | ☐ | ☐ |
| | Understand and use coordinate geometry. | 2002–2004, 2020, 2021 | – | ☐ | ☐ | ☐ |
| | Understand and solve simultaneous equations. | 2005, 2018 | 01S5A, 01S6B | ☐ | ☐ | ☐ |
| | Understand and solve inequalities. | 2008, 2009 | 01S7A | ☐ | ☐ | ☐ |
| **2 Polynomials and the binomial theorem** | Manipulate, simplify and factorise polynomials. | 2006 | 02S1A | ☐ | ☐ | ☐ |
| | Understand, explore and use the binomial theorem. | 2041 | 02S2B | ☐ | ☐ | ☐ |
| | Divide polynomials by algebraic expressions. | 2043 | – | ☐ | ☐ | ☐ |
| | Understand and use the factor theorem. | 2042 | 02S3A | ☐ | ☐ | ☐ |
| | Use a variety of techniques to analyse a function and sketch its graph. | 2022–2024, 2027, 2258 | 02S4B | ☐ | ☐ | ☐ |
| **3 Trigonometry** | Calculate the values of sine, cosine and tangent for angles of any size. | 2047, 2048, 2257 | – | ☐ | ☐ | ☐ |
| | Use the two identities $\sin^2\theta + \cos^2\theta = 1$ and $\tan\theta = \sin\theta \div \cos\theta$ and recognise $x^2 + y^2 = r^2$ as the equation of a circle. | 2053, 2257, 2284 | 03S1A | ☐ | ☐ | ☐ |
| | Sketch graphs of trigonometric functions and describe their main features. | 2047, 2048 | 03S1A | ☐ | ☐ | ☐ |
| | Solve various types of trigonometric equations. | 2047, 2048, 2053, 2257, 2284, 2285 | 03S1A | ☐ | ☐ | ☐ |
| | Use the sine and cosine rules and the area formula for a triangle. | 2045, 2046 | 03S2B | ☐ | ☐ | ☐ |
| **4 Differentiation and integration** | Differentiate from first principles. | 2028 | 04S1A | ☐ | ☐ | ☐ |
| | Differentiate functions composed of terms of the form $ax^n$ | 2029 | 04S2B | ☐ | ☐ | ☐ |
| | Use differentiation to calculate rates of change. | 2269, 2270 | – | ☐ | ☐ | ☐ |
| | Work out equations, tangents and normals. | 2030 | 04S4B | ☐ | ☐ | ☐ |
| | Work out turning points and determine their nature. | 2270 | 04S5A | ☐ | ☐ | ☐ |
| | Work out and interpret the second derivative. | 2270 | 04S5A | ☐ | ☐ | ☐ |
| | Work out the integral of a function. | 2054, 2055 | 04S6B | ☐ | ☐ | ☐ |
| | Understand and calculate definite integrals. Use them to calculate the area under a curve. | 2056, 2273 | 04S7A | ☐ | ☐ | ☐ |

# Objectives checklist

| Ch | Objective | MyMaths | InvisiPen | No | Almost | Yes! |
|---|---|---|---|---|---|---|
| **5 Exponentials and logarithms** | Convert between powers and logarithms. | 2062 | 05S1B | ☐ | ☐ | ☐ |
| | Manipulate expressions and solve equations involving powers and logarithms. | 2062, 2063, 2257 | 05S1B, 05S2A | ☐ | ☐ | ☐ |
| | Use the exponential functions $y = a^x$, $y = e^x$, $y = e^{kx}$ and their graphs. | 2061, 2133, 2134, 2136 | 05S2A, 05S3B | ☐ | ☐ | ☐ |
| | Verify and use mathematical models, including those of the form $y = ax^n$ and $y = kb^x$ | 2268 | 05S4A | ☐ | ☐ | ☐ |
| | Consider limitations of exponential models. | – | – | ☐ | ☐ | ☐ |
| **6 Vectors** | Identify vector quantities and scalar quantities. | 2206 | – | ☐ | ☐ | ☐ |
| | Solve geometric problems in two dimensions using vector methods. | 2206 | 06S1B | ☐ | ☐ | ☐ |
| | Solve problems involving displacements, velocities and forces. | 2206 | – | ☐ | ☐ | ☐ |
| | Find and use the components of a vector. | 2207 | 06S2A | ☐ | ☐ | ☐ |
| | Find the magnitude and direction of a vector expressed in component form. | 2207 | 06S2A | ☐ | ☐ | ☐ |
| | Use position vectors to find displacements and distances. | 2207 | – | ☐ | ☐ | ☐ |
| **7 Units and kinematics** | Understand and use standard SI units and convert between them and other metric units. | 2183 | – | ☐ | ☐ | ☐ |
| | Calculate average speed and average velocity. | 2183 | 07S1B | ☐ | ☐ | ☐ |
| | Draw and interpret graphs of displacement and velocity against time. | 2183 | 07S2A | ☐ | ☐ | ☐ |
| | Derive and use the formulae for motion in a straight line with constant acceleration. | 2184 | 07S3B | ☐ | ☐ | ☐ |
| | Use calculus to solve problems involving variable acceleration. | 2289 | 07S4A | ☐ | ☐ | ☐ |
| **8 Forces and Newton's laws** | Resolve in two perpendicular directions for a particle in equilibrium. | 2186, 2293 | 08S1A | ☐ | ☐ | ☐ |
| | Calculate the magnitude and direction of the resultant force acting on a particle. | 2293 | – | ☐ | ☐ | ☐ |
| | Resolve for a particle moving with constant acceleration. Work out acceleration of forces. | 2187, 2293 | 08S2B | ☐ | ☐ | ☐ |
| | Understand the connection between the mass and the weight of an object. Know that weight changes depending on where the object is. | 2185, 2187 | 08S3A | ☐ | ☐ | ☐ |
| | Resolve for "connected objects", such as an object in a lift. | 2188 | – | ☐ | ☐ | ☐ |
| | Resolve for particles moving with constant acceleration connected by string over pulleys. | 2188 | 08S4B | ☐ | ☐ | ☐ |

| Ch | Objective | MyMaths | InvisiPen | No | Almost | Yes! |
|---|---|---|---|---|---|---|
| **9 Collecting, representing and interpreting data** | Distinguish a population and its parameters from a sample and its statistics. | 2275 | 09S1A | ☐ | ☐ | ☐ |
| | Identify and name sampling methods. | 2275 | 09S1A | ☐ | ☐ | ☐ |
| | Highlight sources of bias in a sampling method. | 2275 | – | ☐ | ☐ | ☐ |
| | Read continuous data given in box-and-whisper plots, histograms and cumulative frequency diagrams. | 2276–2278 | 09S3A | ☐ | ☐ | ☐ |
| | Plot scatter diagrams and use them to identify types and strength of correlation. | 2283 | 09S4B | ☐ | ☐ | ☐ |
| | Use scatter diagrams and rules using quantities to identify outliers. | 2283 | 09S4B | ☐ | ☐ | ☐ |
| | Summarise raw data using appropriate measures of location and spread. | 2279 –2282 | 09S2B | ☐ | ☐ | ☐ |
| **10 Probability and discrete random variables** | Use the vocabulary of probability theory, including the following terms: random experiment, sample space, independent events and mutually exclusive events. | 2093 | 10S1B | ☐ | ☐ | ☐ |
| | Solve the problems involving mutually exclusive and independent events using the addition and multiplication rules. | 2093, 2094 | 10S1B | ☐ | ☐ | ☐ |
| | Use a probability function or a given context to find the probability distribution and probabilities for particular events. | 2114 | – | ☐ | ☐ | ☐ |
| | Recognise and solve problems relating to experiments which can be modelled by the binomial distribution. | 2110, 2111 | 10S2A | ☐ | ☐ | ☐ |
| **11 Hypothesis testing 1** | Understand the terms null hypothesis and alternative hypothesis. | 2115 | 11S1A, 11S2B | ☐ | ☐ | ☐ |
| | Understand the terms critical value, critical region and significance level. | 2115 | 11S1A | ☐ | ☐ | ☐ |
| | Calculate the critical region. | 2115 | 11S1A | ☐ | ☐ | ☐ |
| | Calculate the $p$-value. | 2115 | – | ☐ | ☐ | ☐ |
| | Decide whether to reject or accept the null hypothesis. | 2115 | 11S1A | ☐ | ☐ | ☐ |
| | Make a conclusion based on whether you reject or accept the null hypothesis. | 2115 | 11S2B | ☐ | ☐ | ☐ |

# Answers

## Paper 1 (Set A)

**1 a i** $\dfrac{x^4}{4}+c$

    **ii** $-\dfrac{1}{x}+c$   or   $-x^{-1}+c$

    **iii** $-\dfrac{1}{4x^4}+c$   or   $-\dfrac{1}{4}x^{-4}+c$

  **b** $-3x^{-1}-\dfrac{1}{2}x^4-\dfrac{3}{4}x^{-4}-2x+c$   or

    $-3\dfrac{1}{x}-\dfrac{1}{2}x^4-\dfrac{3}{4}\dfrac{1}{x^4}-2x+c$

**2 a** $f(6)=6^4-6^3-22(6)^2-44(6)-24=0$

    $\therefore x-6$ is a factor

    $f(-1)=(-1)^4-(-1)^3-22(-1)^2-44(-1)-24=0$

    $\therefore x+1$ is a factor

  **b** $f(x)=(x-6)(x+1)(x+2)^2$

**3** $1:\dfrac{11}{5}$ or $1:2.2$

**4** $x=-1+\dfrac{\sqrt{22}}{2}$   or   $x=\dfrac{-2+\sqrt{22}}{2}$

**5** $\log_{10}\sqrt{\dfrac{b}{c}}$

**6** $-\dfrac{7}{8}$

**7** $q'(x)=\lim\limits_{h\to0}\dfrac{q(x+h)-q(x)}{h}$

    $q'(x)=\lim\limits_{h\to0}\dfrac{4(x+h)^2-2(x+h)-(4x^2-2x)}{h}$

    $q'(x)=\lim\limits_{h\to0}\dfrac{4x^2+8xh+4h^2-2x-2h-4x^2+2x}{h}$

    $q'(x)=\lim\limits_{h\to0}(8x+4h-2)$

    $q'(x)=8x-2$

**8 a** Radius = 2, Centre = (3, 6)

  **b** Translation $\begin{pmatrix}3\\6\end{pmatrix}$

  **c i** $y=-x+9$

    **ii** $(3+\sqrt{2},6-\sqrt{2})$ and $(3-\sqrt{2},6+\sqrt{2})$

**9 a** $\theta=0°,90°$

  **b** $\theta=9°,27°,45°,63°,81°$

**10** $26a^3+\dfrac{2}{a}$

**11 a i** $\log_{10}y=\log_{10}(ab^x)$

    $\log_{10}y=\log_{10}a+\log_{10}b^x$

    $\log_{10}y=\log_{10}a+x\log_{10}b$

    $\log_{10}y=x\log_{10}b+\log_{10}a$

    **ii** Points lie on straight line $\Rightarrow$ Assertion correct

  **b** $a=1.6,\ b=1.1$

**12 a i** $y=\dfrac{3}{2}x+\dfrac{1}{2}$

    **ii** Intersection of line and either circle

    $x^2+\left(\dfrac{3}{2}x+\dfrac{1}{2}\right)^2-2x+3\left(\dfrac{3}{2}x+\dfrac{1}{2}\right)=9$

    $x^2+\dfrac{9}{4}x^2+\dfrac{3}{2}x+\dfrac{1}{4}-2x+\dfrac{9}{2}x+\dfrac{3}{2}=9$

    or

    $4(x+2)^2+4\left(\dfrac{3}{2}x\right)^2=45$

    $4x^2+16x+16+9x^2=45$

    $13x^2+16x-29=0$

    $(13x+29)(x-1)=0$

    $x=1,-\dfrac{29}{13}$

$y=\dfrac{3}{2}\times1+\dfrac{1}{2}=2$   or   $y=\dfrac{3}{2}\times-\dfrac{29}{13}+\dfrac{1}{2}=-\dfrac{37}{13}$

Coordinates are $A(1,2)$ and $B\left(-\dfrac{29}{13},-\dfrac{37}{13}\right)$

  **b** 5.824

**13 a** $h=4-7x$

  **b** $2(3x\times h+4x\times h+3x\times4x)$

    $=14hx+24x^2$

    $=14(4-7x)+24x^2$

    $=56x-74x^2$

  **c i** $10\dfrac{22}{37}\ \text{cm}^2$

    **ii** $1\dfrac{13}{37}\ \text{cm}$ by $1\dfrac{5}{37}\ \text{cm}$ by $1\dfrac{19}{37}\ \text{cm}$

## Paper 2 (Set A)

**1 a i** 24.1°C

    **ii** 2.38°C

  **b** Beijing

  **c i** 75.3°F

    **ii** 4.29° F

**2 a** Cluster sampling

  **b** $\dfrac{3}{11}$   or   0.273 (3 sf)

  **c** Knowing the amount of rainfall tells you nothing about the amount of sunshine, and vice versa.

**3 a**

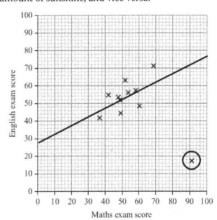

  **b** The point is very different from the rest of the class.

  **c** 57 – 68 marks

  **d**

| | | | |
|---|---|---|---|
| $0\le x<40$ | 1 | $40\le x<45$ | 1 |
| $45\le x<50$ | 3 | $50\le x<55$ | 2 |
| $55\le x<60$ | 1 | $60\le x<70$ | 2 |
| $70\le x<100$ | 1 | | |

  **e i** $\dfrac{6}{11}$   or   0.545 (3 sf)

    **ii** $\dfrac{1}{2}$   or   0.5

**4 a i** Each station is given a number from 1 to 3000…

    …and 18 different random numbers are generated in this range.

    Use the stations corresponding to those numbers.

    **ii** The sample draws at random from the population and any station is equally likely to be chosen.

  **b** Let $p=$ the probability that a station is not fit for sale.

    $H_0: p=0.06$

    $H_1: p>0.06$

    Let $X=$ the number of stations that are unfit for sale.

    $X\sim\text{Bin}(18,0.06)$

    p-value $=P(X\ge3)=0.0898$ (3 sf)

    or

    critical region $X\ge4$

0.0898 > 0.05
or 3 < 4
   There is insufficient evidence to reject the null hypothesis. The probability that a station isn't fit for sale has not increased.

5  37.6 (3 sf)

6  170 m

7  $s = \dfrac{4}{3}t^3 - 3t^2 + 3t + 5$

8  a  $y = \pm 12$ N
   b  $\pm 18.9°$

9  a  63 m
   b  36 m s$^{-1}$
   c  2.71 s (3 sf)
   d  Air resistance will reduce the particle's acceleration. The time taken will increase.

10 a  Newton's second law, applied vertically
      $8g - T = 8a$
      $80 - T = 8a$
   b  $a = 2\dfrac{4}{13}$ m s$^{-2}$ = 2.31 m s$^{-2}$ (3 sf)
      $T = 61\dfrac{7}{13}$  N = 61.5 N  (3 sf)

## Paper 1 (Set B)

1  a  1.4314    b  2.4771    c  1.0458

2  a
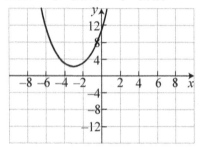

   b
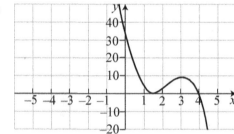

   c
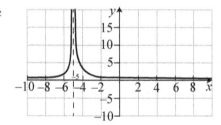

   d
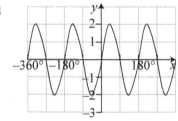

3  $t^2 - t = t(t-1)$
   $t$ even  $\Rightarrow$  $t - 1$ is odd
   $t$ odd  $\Rightarrow$  $t - 1$ is even
   even × odd = even
   odd × even = even
   $\therefore$ $t^2 - t$ is even
   OR
   $t$ even $\Rightarrow$ $t^2$ even
   even − even = even

$t$ odd $\Rightarrow$ $t^2$ odd
odd − odd = even
$\therefore$ $t^2 - t$ is even

4  Linda
   Right-hand side must be zero before factorising.
   OR
   Correct answers are 1 and 5

5  $x = -1, y = 3$ or $x = 9, y = -\dfrac{19}{2}$

6  $\theta = 57.5°, 122.5°, 216.4°, 323.6°$

7  a  $6x^2 + 8x - 4 = p(2x - 1)$
      $6x^2 + (8 - 2p)x - 4 + p = 0$
      $6x^2 + 2(4 - p)x + p - 4 = 0$
   b  i  Discriminant > 0
         $(8 - 2p)^2 - 4 \times 6 \times (p - 4) > 0$
         $64 - 32p + 4p^2 - 24p + 96 > 0$
         $4p^2 - 56p + 160 > 0$
         $p^2 - 14p + 40 > 0$
      ii  $p < 4$ or $p > 10$

8  37.6 cm$^2$

9  a  $0.555\mathbf{i} - 0.832\mathbf{j}$
   b  $\sqrt{17}$, 104°
   c  i  $\overrightarrow{AB} = \overrightarrow{OB} - \overrightarrow{OA} = 2\mathbf{i} - 3\mathbf{j}$
         $\overrightarrow{BC} = \overrightarrow{OC} - \overrightarrow{OB} = 4\mathbf{i} - 6\mathbf{j}$
         $\overrightarrow{BC} = 2\overrightarrow{AB}$
         $\overrightarrow{AB}$ and $\overrightarrow{BC}$ are parallel and both pass through point $B$
         $\therefore$ A, B and C are collinear.
      ii  $\dfrac{2}{3}$

10 a  £750     b  £458.05     c  2.94 years
   d
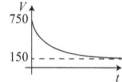

   e  Not a realistic model, as the model suggests that the value of the phone will never be less than £150

11  $(2 - \sqrt{3})$ cm

12 a  $y = -2x^2 + x + 1$     b  $-\dfrac{1}{2}$     c  $\dfrac{9}{8}$

13 a  $\dfrac{du}{dx} = 15x^2 + 8x$     b  $u = 5x^3 + 4x^2 - 1$

## Paper 2 (Set B)

1  a  0.0075
   b  i  0.255
      ii  0.027 (3 sf)

2  a  i  Take data from any members of the population that are available.
      ii  4, 1.31 (3 sf)
   b  i  Not every member of the population is as likely to have an equal chance of being selected using opportunity sampling.
      ii  3.2, 0.705 (3 sf)
      iii The first sample has a higher average.
          The first sample is more spread out around that average.

3  a  Frequencies: 12, 7, 4, 4, 3, 0
   b  $\dfrac{1}{6}$  or  0.167
   c  i  Any two of Fixed number of trials;
         there are 14 days over the two weeks.
         Fixed probability of success;
         it is $\dfrac{1}{6}$ (FT answer to part b).
         There are two possible outcomes;
         $\geq 6$ cm or < 6 cm of rainfall.
         Trials are independent;
         the amount of rainfall on one day shouldn't affect the amount of rainfall on any other day.

**ii** Let $p$ = the probability that 6 cm or more of rainfall occurs on any one day.

$H_0: p = \dfrac{1}{6}$, $H_1: p < \dfrac{1}{6}$

Let $X$ = the number of days when less than 6 cm of rain falls.

$X \sim B(14, \dfrac{1}{6})$

p-value $= P(X = 0) = 0.0779$

$0.0779 < 10\%$

There is sufficient evidence to reject the null hypothesis.
Less rain fell during the sunny weeks.

**4**  12 m

**5**  **a**  Stationary    **b**  $7.5\,\text{ms}^{-1}$

**c**

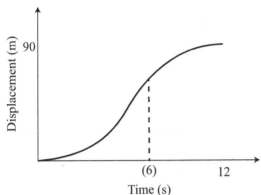

**6**  **a**  $12 - 6t\,\text{ms}^{-2}$

  **b**  $s = 6t^2 - t^3 + c$

    $c = 5$

    $s = 6 \times 4^2 - 4^3 + 5$

    $= 96 - 64 + 5$

    $= 37$

  **c**  $t = 0$ and 4 s

**7**  **a**  $0.15\,\text{ms}^{-2}$    **b**  280 N

**8**  79.9 m (3 sf)

## Paper 1 (Set C)

**1**  **a**  $115 = 5 \times 23$    $\Rightarrow$ not prime

    $116 = 2^2 \times 29$    $\Rightarrow$ not prime

    $117 = 3 \times 39$    $\Rightarrow$ not prime

    $118 = 2 \times 59$    $\Rightarrow$ not prime

    $119 = 7 \times 17$    $\Rightarrow$ not prime

    $120 = 2^3 \times 3 \times 5$  $\Rightarrow$ not prime

    $121 = 11^2$    $\Rightarrow$ not prime

    $122 = 2 \times 61$    $\Rightarrow$ not prime

    $123 = 3 \times 41$    $\Rightarrow$ not prime

    $124 = 2^2 \times 31$    $\Rightarrow$ not prime

    $125 = 5^3$    $\Rightarrow$ not prime

    All numbers have factors (other than 1 and the number itself)

    $\therefore$ The statement is true.

    There are no primes between 115 and 125

  **b**  For example, $a = \sqrt{2}, b = \sqrt{2}$

**2**  **a**  $135°$

  **b**  **i**  $-\dfrac{8}{17}$

    **ii**  $-\dfrac{8}{15}$

  **c**  $\left(\dfrac{\cos x}{1 - \sin x}\right) \cdot \left(\dfrac{1 + \sin x}{1 + \sin x}\right)$

    $= \dfrac{\cos x + \cos x \sin x}{1 - \sin^2 x}$

    $= \dfrac{\cos x + \cos x \sin x}{\cos^2 x}$

    $= \dfrac{1 + \sin x}{\cos x}$

    OR

$\left(\dfrac{\cos x}{1 - \sin x}\right) \cdot \dfrac{\cos x}{\cos x}$

$= \dfrac{1 - \sin^2 x}{(1 - \sin x)\cos x}$

$= \dfrac{(1 - \sin x)(1 + \sin x)}{(1 - \sin x)\cos x}$

$= \dfrac{1 + \sin x}{\cos x}$

**3**  **a**  $-8 < x \le \dfrac{16}{3}$    **b**  $x < \dfrac{17}{10}$

  **c**  No

    For $x < -4$, $(x + 4) < 0$ but did not reverse inequality.

    or

    Correct answer $x < -6$ or $x > -4$

**4**  **a**  He has divided through by $\sin x$ and therefore lost the solutions to $\sin x = 0$

    OR

    Missed out $x = 0°$, $x = 180°$

  **b**  $x = 0°, 19.5°, 160.5°, 180°$

**5**  18 and 20

**6**  **a**  **i**  $\dfrac{dy}{dx} = 12x^3 + 54x^2 - 30x$

    **ii**  $\dfrac{d^2 y}{dx^2} = 36x^2 + 108x - 30$

  **b**  When $x = 1$, $\dfrac{dy}{dx} = 36$

    $\dfrac{dy}{dx} > 0 \Rightarrow$ increasing

  **c**  $(-5, -744)$ is a minimum, $(0, 6)$ is a maximum, $\left(\dfrac{1}{2}, \dfrac{75}{16}\right)$ is a minimum

**7**  **a**  **i**  $\log_{10} y = \log_{10}(kx^n)$

      $\log_{10} y = \log_{10} k + \log_{10} x^n$

      $\log_{10} y = \log_{10} k + n\log_{10} x$

    **ii**  Points lie on a straight line

      $\Rightarrow$ Belief correct

  **b**  $n = 1.6, k = 2.5$

**8**  **a**

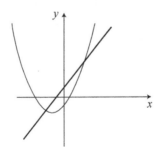

  **b**

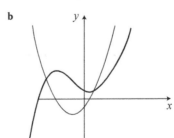

**9**  **a**  28.3

  **b**  **i**  $\mathbf{d} + \mathbf{e} + \mathbf{f}$

    **ii**  $\overrightarrow{OD} + \overrightarrow{OE} + \overrightarrow{OF}$

**10 a i**

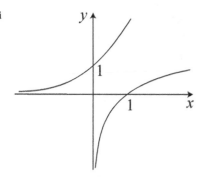

    **ii** They are inverses of one another.
    **iii** Their graphs are reflections of one another in the line $y = x$
  **b** $x = -2.20$

**11 a i** $y = 4x + 5$

    **ii** $y = -\dfrac{1}{2}x + \dfrac{9}{4}$

  **b** $\left(-\dfrac{11}{18}, \dfrac{23}{9}\right)$

  **c** $\left(x + \dfrac{11}{18}\right)^2 + \left(y - \dfrac{23}{9}\right)^2 = \dfrac{7565}{324}$

**12 a** $\dots 56a^5b^3 + 70a^4b^4 + 56a^3b^5 + 28a^2b^6 + 8ab^7 + b^8$
  **b** 0.0390

## Paper 2 (Set C)

**1 a i** $\dfrac{1}{7}$     **ii** $\dfrac{1}{3}$

  **b** $\text{P(red and triangle)} = \dfrac{14}{112} = \dfrac{1}{8}$

    $\text{P(red)} \times \text{P(triangle)} = \dfrac{49}{112} \times \dfrac{40}{112} = \dfrac{5}{32}$

    $\neq \text{P(red and triangle)}$

    $\Rightarrow$ The events are not independent.

  **c**

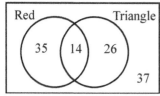

**2 a** 0.0246
  **b** 0.209
  **c** 0.263
  **d** $H_0: p = 0.4\%$, $H_1: p < 0.4$
    $X \sim \text{B}(18, 0.4)$
    p-value $= \text{P}(X \leq 5) = 0.209$
    $0.209 > 0.2$
    Do not reject $H_0$
    There is insufficient evidence to say that the preference has decreased.

**3 a** Stratified
  **b** Min = 1, LQ = 15, Med = 28, UQ = 49, Max = 68
    IQR = 34
  **c** For example
    The office workers liked darker chocolate on average.
    Both groups have the same interquartile range.
    The students have a smaller total range, indicating more of a consensus of opinion.

**4 a** Countryside $\approx 17\%$, Coastal $\approx 24\%$
  **b** They have found the median age.
    Using a line across from 400 people, 50% of 800, and down to 43 years.
  **c** Cluster sampling is used when each cluster is similar to any other.
    The political preferences in the two towns are unlikely to be similar...
    ...because the ages aren't similar.

**5 a** $41.4°$ (3 sf) (upstream)
  **b** 29.9 s (3 sf)

**6** $2.87 \text{ m s}^{-2}$ (3 sf)

**7 a** $53.3 \text{ m s}^{-1}$ (3 sf)
  **b** The acceleration is *not* constant.
  **c** Weight is acting downwards.
    Air resistance is acting upwards.
    The forces are equal and opposite.
    785 N (3 sf)

**8 a** $2.67 \text{ m s}^{-2}$ (3 sf)     **b** 0 s and 25 s     **c** 69 s

**9 a** $v = 3t^2 - 4t + 3$
  **b** $s = t^3 - 2t^2 + 3t + c$
    $0 = 0 + c$
    $s = t^3 - 2t^2 + 3t$
    $0 = t^3 - 2t^2 + 3t$
    $= t(t^2 - 2t + 3)$
    $'b^2 - 4ac' = 2^2 - 4 \times 1 \times 3$
    $= -8$
    $-8 < 0 \Rightarrow$ no real solutions
    Therefore only solution at $t = 0$